"This book serves as an important inventory and source of knowledge about the new approach needed to tackle some of the challenges facing human kind in the present era. The book will be relevant for a very long time as a lesson to scholars and a reference for policy makers."

From the foreword by Albert T Modi, Dean and Head of School of Agricultural, Earth and Environmental Sciences, University of KwaZulu-Natal, South Africa

T0133516

Community Innovations in Sustainable Land Management

It is increasingly recognized that land can be managed most sustainably through involving local communities. This book highlights the potential of a new methodology of uncovering and stimulating community initiatives in sustainable land management in Africa.

Analyses of four contrasting African countries (Ghana, Morocco, South Africa and Uganda) show that, as communities directly face the challenges of land degradation, they are likely to develop initiatives themselves in terms of sustainable land management. These initiatives (or 'innovations') may be more appropriate and sustainable than those emanating from research stations located far from the communities. The book describes the rationale of the approach used, the set of steps followed, how the project managed to engage the communities to understand the importance of the activities they were undertaking and how they were stimulated to improve and extend their initiatives and innovativeness.

Examples covered include soil fertility, community forestry, afforestation, water, invasive species and grazing land management. Central to the book is the way communities, and scientists, interacted between the four countries and learnt from each other. The book also shows how the initiatives were outscaled locally.

Maxwell Mudhara lectures at the School of Agricultural, Earth and Environmental Sciences and is Director of the Farmer Support Group, University of KwaZulu-Natal, South Africa.

William Critchley is a Director of Sustainable Land Management Associates Ltd, UK and former Senior Advisor at the Sustainable Land Management Thematic Unit, Centre for International Cooperation, Vrije Universiteit, Amsterdam, Netherlands.

Sabina Di Prima is a Sustainable Land Management Specialist at the Centre for International Cooperation, Vrije Universiteit, Amsterdam, Netherlands.

Saa Dittoh teaches in the Department of Climate Change and Food Security, University for Development Studies, Tamale, Ghana and was formerly Head of the University's Food and Nutrition Security Unit.

Mohamed F. Sessay is a Senior Programme Officer with the Biodiversity Unit of the Division of Environmental Policy Implementation (DEPI), United Nations Environment Programme (UNEP), Nairobi, Kenya. He was Chief of the Global Environment Facility (GEF) Biodiversity/Land Degradation/Biosafety Unit in DEPI, UNEP until his retirement in March 2015.

Earthscan Studies in Natural Resource Management

For more information on books in the Earthscan Studies in Natural Resource Management series, please visit the series page on the Routledge website: www.routledge.com/books/series/ECNRM.

Community Innovations in Sustainable Land Management

Lessons from the field in Africa

**Edited by Maxwell Mudhara,
William Critchley, Sabina Di Prima,
Saa Dittoh and Mohamed F. Sessay**

Routledge
Taylor & Francis Group
LONDON AND NEW YORK

from Routledge

First published 2016 by Routledge

2 Park Square, Milton Park, Abingdon, Oxfordshire OX14 4RN

711 Third Avenue, New York, NY 10017

Routledge is an imprint of the Taylor & Francis Group, an informa business

First issued in paperback 2018

British Library Cataloguing-in-Publication Data
A catalogue record for this book is available from the British Library

Library of Congress Cataloging in Publication Data
Names: Mudhara, Maxwell, editor.
Title: Community innovations in sustainable land management : lessons from the field in Africa / edited by Maxwell Mudhara, Saa Dittoh, Mohamed Sessay, William Critchley and Sabina Di Prima.
Description: London ; New York : Routledge, 2016. | Series: Earthscan studies in natural resource management | Includes bibliographical references and index.
Identifiers: LCCN 2016001165| ISBN 9781138190474 (hbk) | ISBN 9781315641003 (ebk)
Subjects: LCSH: Land degradation--Control--Africa.
Classification: LCC GE160.A35 C66 2016 | DDC 333.73096--dc23
LC record available at http://lccn.loc.gov/2016001165

ISBN: 978-1-138-19047-4 (hbk)
ISBN: 978-0-367-02970-8 (pbk)

Typeset in Bembo
by Fish Books Ltd.

Contents

List of figures

List of tables

Abbreviations

AGDP	agricultural gross domestic product
ARD	agricultural research and development
BANDERA	Balimi Network for Developing Enterprises in Rural Agriculture (Uganda)
CA	conservation agriculture
CAADP	Comprehensive African Agriculture Development Programme
CBD	Convention on Biological Diversity
CBO	community-based organisation
CBNRM	Community-Based Natural Resource Management (South Africa)
CEAD	Centre for Environment and Development (South Africa)
CEO	chief executive officer
CERDES	Centre for Rural Development Systems (South Africa)
CFM	Community forest management/Collaborative forest management
CFR	Central Forest Reserve (Uganda)
CGIAR	Consultative Group on International Agricultural Research
CI	community initiatives
CIS-VU	Centre for International Cooperation, Vrije Universiteit Amsterdam
CoGTA	Department of Cooperative Governance and Traditional Affairs (South Africa)
COMESA	Common Market for Eastern and Southern Africa
CSIF	Country Strategic Investment Framework (Ghana)
CTA	Technical Centre for Agricultural and Rural Cooperation
CWSSE	Conserve Water to Save the Soil and the Environment (project, Uganda)
DAFF	Department of Agriculture, Forestry and Fisheries (South Africa)
DAO	District Agricultural Officer (Uganda)
DAP	di-ammonium phosphate
DEA	Department of Environmental Affairs (South Africa)

DEAT	Department of Environmental Affairs and Tourism (South Africa)
DFID	Department for International Development (UK)
DGIA	Directorate General for International Cooperation (Netherlands)
DRDLR	Department of Rural Development and Land Reform (South Africa)
DSIP	Agricultural Sector Development Strategy and Investment Plan (Uganda)
DWF	Department of Water and Forestry (Morocco)
FAIR	Farmer Access to Innovation Resources (project)
FAO	Food and Agriculture Organization
FFS	Farmer Field Schools
FSG	Farmer Support Group (South Africa)
GDP	gross domestic product
GEAR	Growth, Employment and Redistribution (programme; South Africa)
GEBs	global environmental benefits
GEF	Global Environment Facility
GFAR	Global Forum on Agricultural Research
HEIA	high external input agriculture
IC	innovative community
ICOUR	Irrigation Company of Upper Region (Ghana)
IFAD	International Fund for Agricultural Development
IGAD	Intergovernmental Authority on Development (Uganda)
IK	indigenous knowledge
ISRDS	Integrated Sustainable Rural Development Strategy (South Africa)
ISWC	Indigenous Soil and Water Conservation (project, Uganda)
LDFA	land degradation focal area
LEIA	low external input agriculture
LEISA	low external input and sustainable agriculture
M&E	monitoring and evaluation
MAAIF	Ministry of Agriculture, Animal Industry and Fisheries (Uganda)
MDGs	Millennium Development Goals
MEST	Ministry of Environment, Science and Technology (Ghana)
MLNR	Ministry of Lands and Natural Resources (Ghana)
MOFA	Ministry of Food and Agriculture (Ghana)
MSP-B	medium sized project brief
MTADP	Medium Term Agricultural Development Programme (Ghana)
NAADS	National Agricultural Advisory Services (Uganda)
NACIA	Nalukonge Community Initiative Association (Uganda)
NAP	National Action Plan (Morocco)
NAP	National Action Programme (under UNCCD)

NARLI	National Agriculture Research Laboratories Institute (Uganda)
NARO	National Agriculture Research Organisation (Uganda)
NARP	National Agricultural Research Policy (Uganda)
NDA	National Department of Agriculture (South Africa)
NDP	National Development Plan (Uganda)
NEAP	National Environmental Action Plan (Ghana; Uganda)
NEIA	no external input agriculture
NEMA	National Environment Management Act (South Africa)
NEMA	National Environment Management Authority (Uganda)
NFA	National Forestry Authority (Uganda)
NGO	non-governmental organisation
NSBCP	Northern Savannah Biodiversity Conservation Project (Ghana)
NSC	National Steering Committee
NYE	Natal Timber Extract
PEAP	Poverty Eradication Action Plan (Uganda)
PELUM	Participatory Ecological Land Use Management
PIF	project identification form
PFI	Promoting Farmer Innovation (project)
PFM	Participatory Forest Management (South Africa)
PID	participatory innovation development
PMA	Plan for the Modernization of Agriculture (Uganda)
PTD	participatory technology development
PROLINNOVA	Promoting local innovation in ecologically-oriented agriculture and natural resource management (programme)
RDP	Reconstruction and Development Programme (South Africa)
R&E	research and extension
RECPA	Rwoho Environment Conservation and Protection Association (Uganda)
REDS	Rural Enterprise Development Services
RFM	Rural Forest Management (South Africa)
RSC	Regional Steering Committee
SAPS	South African Police Services
SCI-SLM	Stimulating Community Initiatives in Sustainable Land Management (project)
SDGs	Sustainable Development Goals
SFM	sustainable forest management
SIF/SLM	Strategic Investment Framework for Sustainable Land Management (Uganda)
SIP	Strategic Investment Program/Partnership (of GEF)
SLM	sustainable land management
SO2	strategic objective 2 (of GEF-LDFA)
SOC	soil organic carbon
SRI (test)	Sustainability, Replicability and Inclusiveness
SRMP	Savannah Resources Management Programme (Ghana)

SWC	soil and water conservation
TAAM	average annual growth rate
TAG	technical advisory group
TEES (test)	Technically effective, Economically valid, Environmentally friendly, Socially acceptable
UDS	University for Development Studies (Ghana)
UKZN	University of KwaZulu-Natal
ULC	UKhahlamba Livestock Cooperative
UNCCD	United Nations Convention to Combat Desertification
UNDP	United Nations Development Programme
UNEP	United Nations Environment Programme
UNFCCC	United Nations Framework Convention on Climate Change
UNSO	United Nations Office to Combat Desertification and Drought
WDCD	World Day to Combat Desertification
WfW	Working for Water (South Africa)
WOCAT	World Overview of Conservation Approaches and Technologies
ZEFP	Zasilari Ecological Farms Project (Ghana)

Contributors

Akuriba, Margaret A. University for Development Studies, Department of Agribusiness Management and Finance, Tamale, Ghana.

Critchley, William Sustainable Land Management Associates Ltd, Pitlochry, UK.

Di Prima, Sabina Vrije Universiteit Amsterdam, Centre for International Cooperation (CIS-VU), Amsterdam, Netherlands.

Dittoh, Saa University for Development Studies, Department of Climate Change and Food Security, Tamale, Ghana.

Mahdi, Mohammed Faculty Member, Member of Association Targa-aide, National Agricultural School, Meknes, Morocco.

Malinga, Michael Agricultural Coordinator, Zimele, Sinodale Centre, Pietermaritzburg, South Africa.

Molo, Richard Researcher and Head, Biological Control Unit, National Agricultural Research Organization (NARO), Uganda.

Mudhara, Maxwell Director, Farmer Support Group, University of KwaZulu-Natal, Scottsville, Pietermaritzburg, South Africa.

Mugerwa, Swidiq Researcher and Head, Livestock Nutrition Programme, National Agricultural Research Organization (NARO), Uganda.

Muwaya, Stephen Program Coordinator: Sustainable Land Management/ UNCCD Focal Point, Ministry of Agriculture, Animal Industry and Fisheries (MAAIF) Uganda.

Nabilse, Cuthbert Kaba University for Development Studies, Department of Climate Change and Food Security, Tamale, Ghana.

Sahadeva, Avrashka University of KwaZulu-Natal, Farmer Support Group, Scottsville, Pietermaritzburg, South Africa.

Sessay, Mohamed F. is a Senior Programme Officer with the Biodiversity Unit of the Division of Environmental Policy Implementation (DEPI), United Nations Environment Programme (UNEP), Nairobi, Kenya. He was Chief of the Global Environment Facility (GEF) Biodiversity/Land Degradation/Biosafety Unit in DEPI, UNEP until his retirement in March 2015.

Shezi, Zanele Project Facilitator, Institute of Natural Resources NPC, Pietermaritzburg, South Africa.

Ssendawula, John Sustainable Land Management Specialist, Ministry of Agriculture, Animal Industry and Fisheries (MAAIF), Uganda.

Tamim, Mohamed Faculty Member, Member of Association Targa-aide, National Institute for Urban and Territorial Planning, Rabat, Morocco.

Tijani, Zakariaa Development Agent, Association Targa-aide, Rabat, Morocco.

Tuijp, Wendelien Vrije Universiteit Amsterdam, Centre for International Cooperation (CIS-VU), Amsterdam, Netherlands.

Weobong, Conrad A. University for Development Studies, Faculty of Renewable Natural Resources, Tamale, Ghana.

Preface

This book brings together both the experience and the enthusiasm underlying the implementation of the Stimulating Community Initiatives in Sustainable Land Management (SCI-SLM) project. We have tried to capture, as best we could, accounts of how the project was rolled out across the participating countries and the take-home lessons from the project as a whole. Having implemented a novel approach to building on and achieving sustainable land management (SLM) at community level – and being convinced that a broader audience could learn from the experience – the stakeholders in the project felt it worthwhile to write a book. Thus, this volume is not only a record of what has transpired, but also a resource for development agents, government and communities, in efforts to identify, support and stimulate community-level initiatives in SLM.

The book sets out the rationale for implementing the project; and it also shows how the project was linked to programmes at national and global levels. The countries presented different opportunities and challenges, which make the experiences rich through the heterogeneous circumstances across the African continent. The methodology used in the project is presented, followed by accounts of the country-level experiences and lessons. The lessons from the cross-learning and methodology precede the conclusions and general recommendations. The teams involved in implementing the project at the country level wrote the country case studies. Other chapters were written by members of the Technical Advisory Group in collaboration with key members of the project.

On behalf of the whole team I would like to acknowledge the significant role played by the community members who participated in the project across the four countries. I also acknowledge the Global Environment Facility (GEF) for funding the work, the United Nations Environment Programme (UNEP) for acting as the implementing agency, the University of KwaZulu-Natal (UKZN) for coordinating the project and the Vrije Universiteit Amsterdam for providing advisory services. I acknowledge Ms Lerato Phali in her dedication to the compilation of the final version of the book.

Maxwell Mudhara
Project Coordinator
UKZN
South Africa

Foreword

There is no question that knowledge about sustainable land management (SLM) has always been at the forefront of human development. One of the major drivers of human progress – that is linked to understanding and manipulation of the surrounding environment – is agriculture. This has led to the sustenance and growth of the human population. While agriculture underpinned the development of civilisation, the accompanying population growth was maintained at the cost of environmental degradation.

As the new knowledge era increased in its impact on the world, creating globalisation, a large portion of the human population started to face the challenge of being less and less reliant on traditional, local knowledge. Globalisation presented challenges associated with a top-down approach to human capacity and community development. This approach was largely characterised by huge investments by governments and other institutional development agents at the expense of locally generated solutions. It was also characterised by deployment of human resources that were not well tuned to challenges on the ground, where communities who needed solutions were located. The approach was to implement short-term, top-down projects, with a view to ensuring that specific agendas were being implemented.

Very few projects realised the importance of developing programmes for the communities with the communities, and then to use that information to develop lessons for broader use by other groups, and onwards to inform development policies. Stimulating Community Initiatives in Sustainable Land Management is one of those rare projects that use a participatory approach, whereby the local knowledge of communities has been seriously considered and documented for the benefit of communities across many countries, with common challenges of environmental management linked to land degradation. It is an exemplary programme where there is generation of knowledge about environmental management in four countries in Sub-Saharan Africa, namely Ghana, Morocco, South Africa and Uganda. The partnerships of community members and researchers involved in the programme showed that local communities involved in seeking ways of achieving SLM can contribute to environmental science and coordinate it for sharing among communities from the four countries.

The programme resonates very well with the Millennium Development Goals (MDGs) of the United Nations. In addition, there is clear evidence that, although the programme was completed before the launch of the Sustainable Development Goals (SDGs), it is indeed an excellent example of how to respond to the SDGs. The key and unique features of the programme include community-initiated innovations in SLM; cross-visits between communities; South–South learning in terms of the regional experience and local knowledge; and a good interdisciplinary contribution of research for community development and relevant policy.

The book is aimed at development practitioners working on SLM at the community level, particularly under semi-arid conditions. It can also be used by those who are interested in participatory approaches and development pathways. Land degradation is an important area of focus whose practice faces the challenge of unavailability of methodologies and steps for implementing them at community level, or of examples showing how successful methods have been employed. This book serves as an important inventory and source of knowledge about the new approach needed to tackle some of the challenges facing humankind in the present era. The book will be relevant for a very long time as a lesson to scholars and a reference for policymakers.

Professor Albert T. Modi
Dean and Head of School
School of Agricultural, Earth and Environmental Sciences
University of KwaZulu-Natal

1 Stimulating community initiatives in sustainable land management

An introduction

William Critchley, Maxwell Mudhara and Mohamed F. Sessay

Introduction

Stimulating Community Initiatives in Sustainable Land Management (SCI-SLM) is a project that focuses on identifying innovative forms of land management amongst communities in four countries in Africa: Ghana, Morocco, South Africa and Uganda. The premise of the project is that there are local community innovations succeeding in combating desertification where formal research recommendations have often failed. The common denominator is initiatives, regarding land, water or plant resources, which have emanated from the communities themselves, demonstrating their capacity to come up with solutions to problems of land degradation internally. SCI-SLM endeavours to help add value to these initiatives – through research partnerships – as well as stimulating these communities to go forward with their efforts. SCI-SLM is documenting the initiatives and encouraging other communities to learn from these focal points through, amongst other ways, cross-visits. Thus cross-learning between communities – and between countries – is a key element. Establishing flow lines of communication about successful initiatives and 'innovativeness' is a central issue. At a higher level, SCI-SLM seeks to institutionalise the concept and mechanisms of such an approach, in relevant government ministries and other national organisations.

The budget is modest: US $2m. Half is from the Global Environment Facility, the remainder is co-finance from participating institutions. SCI-SLM falls under the United Nations Environment Programme as the GEF implementing agency. UNEP is centrally situated in Nairobi, Kenya. The project is coordinated and executed by the Centre for Environment and Development (CEAD) at the University of KwaZulu-Natal in South Africa, which also manages the South African country programme. The other three country programmes are managed in Ghana by the University of Development Studies in Tamale, in Morocco by Targa-Aide (a non-governmental organisation – NGO – with links to the University of Hassan II) and in Uganda by the Ministry of Agriculture, Animal Industry and Fisheries. The Centre for Development Cooperation of the Vrije Universiteit Amsterdam provides technical assistance as the 'advisory group' or 'TAG'. This support is especially targeted at methodology and its development. SCI-SLM is supported and guided by a Steering Committee that meets every year.

SCI-SLM as a Global Environment Facility project – the implications

SCI-SLM was targeted at the GEF because of its dedicated focus on combating land degradation through sustainable land management. The GEF's land degradation focal area (LDFA) came into being around the time of project formulation and, with SCI-SLM's emphasis on upscaling, this made a perfect match with GEF-LDFA. The project document states:

> The project fits with the new GEF-4 land degradation focal area strategy and will contribute to its strategic objective 2 (SO2) on 'upscaling of sustainable land management investments that generate mutual benefits for the global environment and local livelihoods'. The project will contribute to improving and sustaining the economic well-being of people and the preservation and/or restoration of ecosystem functions and services under different socio-economic conditions. SCI-SLM further emphasises partnerships with small farmers – as part of communities – to identify and demonstrate, under field conditions, environmentally friendly and socio-economically viable land management practices that will enhance soil fertility and make more effective use of water. The activities of the project will primarily be carried out by recipient country research institutions and will be up-scaled within the four pilot countries. The project will also support the LDFA's strategic programme no. 1 'Supporting Sustainable Agriculture and Rangeland Management' and LDFA's strategic programme no. 3 'Investing in New and Innovative Approaches to SLM'.

GEF finance under the LDFA requires that, alongside immediate local benefits of improved productivity, global environmental benefits (GEBs) are delivered, especially carbon sequestered in the land – in the soil and in living vegetation – through sustainable land management. Fortunately sustainable land management automatically ensures the generation of GEBs alongside local benefits: good SLM leads to a build-up in soil organic matter and greater primary productivity. Thus carbon is captured in the land, securing better and more stable yields: production systems are made more resilient. Simultaneously carbon dioxide in the atmosphere is reduced.

What differentiates SCI-SLM from most other projects in GEF's land degradation focal area is its focus on local innovation – but simultaneously the regional dimension. The project that is most closely related to SCI-SLM in this respect is the Kagera-Transboundary Agro-ecosystem Management Project (Kagera-TAMP), which has FAO as its implementing and executing agency. Covering the four countries of Burundi, Rwanda, Tanzania and Uganda, Kagera-TAMP has a similar history to SCI-SLM in that it took several years before final approval, as a result of the reorganisation of GEF and its funding procedures (see section below). However Kagera-TAMP differs from SCI-SLM in that the four countries share a river basin (the Kagera) and there are expected to be synergies generated as a result of actions in adjacent countries. The regional dimension of SCI-SLM is quite different. Here the rationale for spanning four countries was that lessons could be shared between nations in North, South, East and West Africa. The suitability of the approach in these different parts of Africa would, it was proposed, give a good indication of its applicability across various environments within Africa.

A brief timeline of SCI-SLM

Stimulating Community Initiatives in Sustainable Land Management was one of a number of projects whose approval – and start-up – was delayed considerably by a reorganisation of the Global Environment Facility in the first decade of this century. First conceived in 2002, the proposed project fell into the GEF pipeline in 2003, and a project preparation facility was made available to develop the project document. A series of 'mobile workshops' were conceived – to bring together a small group in each country – to tour various potential locations and develop the project partially *en route* and then to polish up the findings on return. This project development 'roadshow' worked well and drew up a project proposal the following year. After approval by the GEF council in 2006 for funding under GEF-3, the process stumbled. Reorganisation of the GEF and the GEF's procedures meant that the project had, effectively, to be resubmitted in a different form for funding under GEF-4. It took until 2009 for SCI-SLM to become a reality. While other project proposals foundered and fell by the wayside under this period, SCI-SLM survived thanks to the dedication and determination of its proponents. A timeline is presented below.

2013   Project's first phase draws to a close.
2013   Regional Steering Committee and exchange visit: South Africa.
2012   Mid-term evaluation.
2012   Regional Steering Committee and exchange visit: Ghana.
2011   Regional Steering Committee and exchange visit: Uganda.
2010   Regional Steering Committee and exchange visit: Morocco.
2009   Start-up with inception workshop in South Africa and field activities.
2009   Contract signed (CEAD/UNEP–GEF).
2009   UNEP's approval and then chief executive officer (CEO) endorsement.
2009   Co-financing secured.
2008   PIF endorsement.
2007   PIF submission.
2007   SIP approval.
2006   Intended launch of SCI-SLM but delays due to GEF reorganisation.
2006   Approval by GEF council.
2005   Resubmission of the medium sized project brief (MSP-B) after comments received.
2004   MSP-B completed and submitted to UNEP and GEF.
2003   Mobile reconnaissance missions to the four SCI-SLM countries.
2002   Initial concept drawn up and discussed between Vrije Universiteit Amsterdam, Centre for International Cooperation (CIS-VU) and UNEP.

SCI-SLM: goal and objectives

SCI-SLM was approved as a project under GEF's LDFA with the following overall objective:

> To refine ways of stimulating the further improvement and spread of community-based SLM initiatives, while simultaneously developing a methodology to upscale and institutionally embed SCI-SLM approaches at local and regional level in four pilot countries in Africa. The project will contribute to the Strategic Investment Program's Development and Global Environment Objectives in terms of implementation of policies and on-the-ground investments towards upscaling SLM aligned with national and SIP priorities and reduction of impacts of land degradation on ecosystem functions and services in SIP investment areas. South-to-South exchange and learning between strategically positioned countries is a key element of project design.

Its four components were articulated as follows:

1   identification and analysis of community initiatives in SLM (including monitoring and evaluation – M&E);
2   stimulation and upscaling of community initiatives;
3   awareness raising amongst policymakers;
4   development of methodology for upscaling and institutionally embedding SLM initiatives.

It is also important to point out that SCI-SLM formulated guides to indicate what it precisely meant by *a community, an initiative* and *upscaling*. These were points for discussion at the inception workshop in 2009. That workshop identified key elements/criteria to define what these concepts meant *in the context of the SCI-SLM project* (see also Chapter 3). These are as shown in Table 1.1.

This book: the rationale, the structure and the process

Why this book?

The layout of this book was initially developed at the Regional Steering Committee in Ghana, during 2012. In early 2013, a 'write-shop' was convened in

Table 1.1 Elements of key concepts as used under SCI-SLM

Community	Initiative (= innovation) in SLM
• Common interest group	• New in local terms
• Common values/goals	• Developed by a local community/group
• Common identity	• Little or no help/money from outside
• Defined by the innovation	• Technically and/or socially innovative
	• Have potential to spread

South Africa to come to consensus about content and to begin drafting. The chapters were then formulated in the order that follows here. The logic for the order was that the stage should be set out initially; the individual country experiences would then be documented; and following this, thematic areas would be analysed. Finally, a chapter would draw together more general lessons for programmes working with communities – and especially community innovation – in Sub-Saharan Africa. Some comments follow on each of the chapters.

Chapter 2: local innovation – theory, experience and the basis for SCI-SLM

A brief literature review places SCI-SLM in context. Thus, Chambers (1983) is acknowledged for his key role in showing the potential of participatory approaches and Richards (1985) for recognising the power of indigenous knowledge in agriculture. Reij *et al.* (1996) led the move towards a more systematic study of indigenous systems in SLM, and Critchley *et al.* (1999) showed how innovative farmers could be encouraged to be further creative in East Africa. By the time Velasquez *et al.* (2005) and Sanginga *et al.* (2009) wrote about local innovation and community-led development in the Asia-Pacific zone and Africa respectively, the concepts were beginning to be broadly accepted as valid in terms of development. The theory behind farmer innovation is introduced in this chapter. Five 'fundamentals' of the theory are identified: these are (i) local innovation as the dynamic that has shaped tradition; (ii) innovative farmers as perpetual experimenters; (iii) innovators open to sharing of ideas; (iv) pathways of sharing – particular ways in which and reasons why knowledge is exchanged; (v) despite Western science, local innovation endures. There is also an introduction to the precursors to SCI-SLM: the two projects, Indigenous Soil and Water Conservation Phase II and Promoting Farmer Innovation, and a parallel programme entitled PROLINNOVA (Promoting local innovation in ecologically-oriented agriculture and natural resource management). The emergence of SCI-SLM is traced.

Chapter 3: methodology

The authors of this chapter are drawn from the TAG – the technical advisory group – and they explain how the methodology underpinning SCI-SLM evolved from an earlier project that worked with farmer innovators. The rationale for modifying the methodology is set out: the basic difference being that, while the earlier project worked with individual farmers, SCI-SLM works with communities. Thus the basic approach is similar, but it needs to be modified to accommodate all those within a group who have a stake in the innovation. Methodological frameworks based on steps and flows are presented graphically. The chapter highlights the fact that the refinement of methodology is one of the objectives of SCI-SLM and will form an important output of the overall project. Chapter 11 then looks at the refinement process that has occurred.

Chapter 4: SCI-SLM – innovation begins at programme level

This chapter shows that SCI-SLM is aligned to other national and global initiatives and conventions that seek to address land degradation. For example, all four countries, Ghana, Morocco, South Africa and Uganda are signatories to the Convention on Biological Diversity (CBD), United Nations Convention to Combat Desertification (UNCCD) and the Kyoto Climate Change Convention. The project was informed by the mounting pressure on the need to address the challenges of land degradation with its close connection to climate change. Therefore, the chapter illustrates how the approach can make a significant contribution to finding SLM solutions, a crucial issue in terms of both the environment and rural livelihoods.

Chapter 5: Ghana

The four SLM innovations that were identified in rural locations in northern Ghana are discussed here. The innovations had both technical and social components. They covered community forest conservation, non-burning of crop residues on farm lands and soil fertility management through different composting methods. Furthermore, the chapter traces the modalities for engaging the communities and how communities were stimulated to improve upon the innovations and to cross-learn from other communities. The SCI-SLM project intervention also helped to document the practices, attach some 'scientific', economic and environmental significance to the practices and to draw attention to the practices. The modest 'horizontal upscaling' (spread of practices) achieved through the project is discussed, as well as the 'vertical upscaling' (institutionalisation), showing how the approach was mainstreamed into governmental and non-governmental organisations.

Chapter 6: Morocco

Here, four community initiatives are also presented, demonstrating the innovative capacity and creativity of local communities in Morocco. The community initiatives identified and characterised in the mountain communities of the High Atlas are: (i) modifying ancestral rules of sharing flows to increase water use efficiency, (ii) marketing artisanal woodwork products through a website and establishing rules to share the revenue, alongside assignment of resources for reforestation with fruit as well as timber trees, (iii) organising of carpentry activity through a marketing cooperative, (iv) creating of terraced and irrigated land through progressive rehabilitation of wasteland. In addition, evidence is presented of how the project stimulated and enhanced the initiatives. In-country exchange visits were an integral part of the project and are clearly described for the Morocco experience. Evidence of the effect of these visits on the spread of initiatives is illustrated.

Chapter 7: South Africa

The chapter looks at the four community initiatives identified within two district municipalities in KwaZulu-Natal, uThukela and uMzinyathi. The initiatives include two based on the exotic wattle tree – normally considered an alien invasive, but harnessed for its local benefits by two communities – a third on indigenous forest management and the last on self-imposed communal grazing land rules and regulations. The chapter explores how the project engaged with the communities and helped them to understand their own initiatives, gave them better insight into how they contribute towards effective SLM and in addition helped them to enhance their livelihoods. How the project collaborated with different government sectors and thus allowed communities to understand policies and improve their technical capacity is presented. The systematic spread of the community initiatives to neighbouring communities is documented.

Chapter 8: Uganda

Beginning with a description of the situation regarding land degradation in Uganda, the stage is set to highlight the importance and relevance of the SCI-SLM project for the country. The discussion shows that the project evolved into an integral part of various national SLM initiatives through 'vertical upscaling'. Three community initiatives were stimulated under the project in Uganda. One initiative explored the use of arboreal termites to control terrestrial termites and simultaneously set up a *kraaling* system (enclosing cattle at night) to rehabilitate degraded rangelands. The second initiative involved setting up, demonstrating and upscaling SLM practices to improve soil productivity: these included conservation agriculture. The third initiative focused on building a community-wide movement to change the land use of bare, degraded, steep hills from communal livestock grazing to tree planting, through awareness raising, training, negotiations, bylaws and mobilisation of partnerships to ensure continuous and long-term soil protection within the fragile ecosystem. Evidence of the role of the approach in enhancing the interaction between communities, researchers and policy practitioners in a mutual learning undertaking is demonstrated.

Chapter 9: cross-learning with community initiatives

Cross-learning, that is mutually beneficial exchange of experience, was a cornerstone of the SCI-SLM methodology. The chapter demonstrates how the regional approach, which included community initiatives practised in the four corners of Africa, offered a unique opportunity for cross-learning experiences. The inter-country visits demonstrated that exchange of experience between countries can enable communities to learn from each other – based on approaching similar problems using ideas and insights from abroad. Evidence presented shows that the cross-visits between the communities within countries created an important platform for farmer-to-farmer learning experiences, exchange of valuable knowledge and sharing of ideas for all participants, including researchers across all countries. One important impact of the exchange visits was enhancement of farmer confidence to innovate and to be aware of the value of their innovations.

Chapter 10: linking local SLM initiatives to global environmental benefits

This chapter focuses on global environmental benefits of the SCI-SLM interventions emanating from the project across the four participating countries. The evidence showing the global environmental benefits that the innovations contribute, is discussed. The chapter looks at benefits in their broadest global sense, including carbon sequestration, soil fertility improvement, forest improvement, crop and rangelands rehabilitation and, last, livelihoods, gender and poverty reduction – in the context of GEBs.

Chapter 11: evolution of SCI-SLM methodology and approach at programme and country level

In Chapter 3, SCI-SLM's methodology was introduced – it was a modification of what had been successful under a prior farmer innovation project. Under SCI-SLM it was planned to develop this methodology in practice, learning from experience. Each country reflects on how it has used the methodology and what changes it made in practice. The authors of the chapter also bring into account experience at the overall programme level. While it is evident that the country programmes have spent relatively more time on implementation than theorising over methodological steps, several clear lessons emerge. These include the value of the 'TEES' and 'SRI' tests developed for screening projects, and the value of cross-visits as vehicles for upscaling. The TEES-test is used to determine whether a technical innovation has true potential and may lead to long-term/durable benefits and incorporates the assessment of technical, economic, environmental and social benefits of an initiative. The SRI was set up to assess social initiatives with respect to sustainability, replicability and inclusiveness. There is, furthermore, a summary of a research project carried out under SCI-SLM on social innovation which proposed an improvement to the SRI-test.

Chapter 12: lessons learnt and way forward

The chapter draws the lessons from the overall project. The lessons are at different levels, that is, identification of community initiatives, their improvement and upscaling – both spread and institutionalisation. This chapter also illustrates some of the experiences and lessons that emanate from the differences across the communities and the countries. A key lesson is that identification of community initiatives should be systematic, guided by well-defined criteria. Initiatives were improved using insights from a variety of sources such as experiences from other areas, literature and farmers' knowledge. Upscaling amongst groups involved getting the communities to explain and demonstrate their practices, and extending them to others. It emerged that exchange visits within countries, and between them too, were an excellent vehicle to enable communities to learn from each other and to stimulate further innovativeness. Mainstreaming the approach into national agencies and policies emerged as a significant challenge, but one that could be addressed successfully to various extents. It was clear that the circumstances prevailing in each country dictated the manner in which mainstreaming could be achieved. Therefore, flexibility and being opportunistic – based on a clear methodology – could be vehicles for fostering national mainstreaming of the approach.

References

Chambers, R. (1983), *Rural Development. Putting the Poor First*. Longman, London.

Critchley, W., Cooke, R., Jallow, T., Lafleur, S., Laman, M., Njoroge, J., Nyagah, V. and Saint-Firmain, E. (1999), *Promoting Farmer Innovation. Harnessing local environmental knowledge in East Africa*. RELMA and UNDP, Nairobi, Kenya.

Reij, C., Scoones, I. and Toulmin, C. (1996), *Sustaining the Soil*. Earthscan, London.

Richards, P. (1985), *Indigenous Agricultural Revolution*. Hutchinson, London.

Sanginga, P., Waters-Bayer, A., Kaaria, S., Njuki, J. and Wettasinha, C. (eds) (2009), *Innovation Africa – Enriching farmers' livelihoods*. Earthscan, London.

Velasquez, J., Yashiro, M., Yoshimura, S. and Ono, I., (2005), *Innovative Communities – People-centred Approaches to Environmental Management in the Asia-Pacific Region*. United Nations University Press.

2 Local innovation

Theory, experience and the basis for
SCI–SLM

William Critchley and Sabina Di Prima

The theory underpinning local innovation has been well described in previous publications, notably *Promoting Farmer Innovation* (Critchley *et al.* 1999), *Farmer Innovation in Africa* (Reij and Water-Baye 2001), *Working with Farmer Innovators* (Critchley 2007) and Innovation Africa (Sanginga *et al.* 2009). This chapter draws principally on these sources. Local innovation should be recognised and acknowledged as the origin of agricultural development and systems of resource management: it has led to the establishment of traditions. But equally important, it still exists – even though it has been overshadowed by Western science. This creative force can be encouraged and supported to the betterment of livelihoods and the environment: this is underpinned by the experience of previous projects. While proven at individual and family level, it was hypothesised that community innovation would also be vibrant, overlooked and responsive to stimulation (UNEP 2009; Di Prima *et al.* 2013).

The theory of local innovation

Interest in local innovation is a direct consequence of the attention given to participatory processes and indigenous knowledge (IK) in natural resource management that emerged in the 1980s (e.g. Chambers 1983; Millington 1985; Richards 1985; Reij *et al.* 1986). Recognising the importance and noting the potential contribution of IK and participation was the first step: the next was turning it into practical guides on how these could be used in development. Thus, for example, Pretty and colleagues produced *A Trainer's Guide for Participatory Learning and Action* in 1995 (Pretty *et al.* 1995) and Grenier wrote her handbook entitled *Working with Indigenous Knowledge* in 1998 (Grenier 1998). More recently, Pyburn and Woodhill (2014) have produced a primer on the dynamics of rural innovation, which focuses on the growing recognition of 'innovation systems', where 'value chains'– and innovation along those chains – are seen as central processes. While they do not stress the role of local innovation per se, they acknowledge the broad span of actors who construct an innovation system in its entirety.

Local (or 'farmer') innovation emerges from these broad themes as an area with particular potential, but one that can only be fully developed if it is specifically nurtured. Its importance is summarised in the following extract:

Since the time that agriculture began, some 10,000 years ago, it has been shaped and spread through farmers themselves. For practically the whole of that period there was no 'scientific' research. And there were no advisory or extension agencies. Farmers came up with ideas, carried out experiments and arrived at their own conclusions. Innovations that proved to be effective thrived. Crops and animals were selected and bred: tools were made and modified; farming systems were developed. And farmers learned innovations from each other. They bought and bartered improved crop seeds and better livestock from the breeders, and they copied farming systems that worked well. Travel was a great teacher – and the market place a focal point for lively exchange of ideas. In this way traditions were developed and spread. Indeed it could be said that innovation was the dynamic process that led to the development of traditions in farming. Those 'small steps of innovation'… have continuously improved farming practice over time, and still do so now.

(Critchley 2007:1)

The key principles of a local innovation theory are captured in the above. The first is the sheer history of local creativity and experimentation, and the way that it has gradually shaped agriculture – and resource management – over thousands of years. The second point is that farmers are not afraid to experiment. They can be skilful interpreters of results and adapt new ideas to their own specific situation, often turning an enemy into an ally (e.g. an erosion gully into a gully garden; invasive species into productive resources). Third, farmers share – they give and take ideas to and from one another. Fourth, they tend to learn from each other in a particular manner: typically through meetings at the market place, by observation and by visiting each other. The fifth key concept in the above is that this creativity has not been diminished, but merely obscured by Western science: it endures. These five concepts will be now be explored in turn.

History: local innovation has shaped tradition

Indigenous knowledge is often equated to static tradition. In fact much tradition is under continuous evolution – gradually taking new shape through the influence of local innovation. Thus local innovation can be considered as a dynamic form of indigenous knowledge that drives development. 'Small steps' of local innovation have improved technological practices, as well as organisational systems and approaches, over millennia and, despite some technological leaps with the advent of Western scientific research in the nineteenth and twentieth centuries, these small steps continue to have an influence. Figure 2.1 visualises this process.

Farmers experiment

Not all farmers are researchers, just as only some are innovators. But many farmers do test and try variations in systems of production. Others deliberately select plants to improve crop varieties, and others breed animals for different traits. Once again this process has continued since animals were first domesticated and crops were

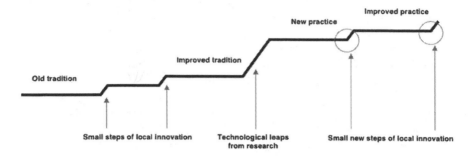

Figure 2.1   Small steps of farmer innovation

selected from wild species. Simultaneously, social arrangements have been established over the millennia: reciprocal working groups and marketing mechanisms for example. These processes persist today, just as they have done for thousands of years. The advent of scientific research has given a huge boost to productivity by accelerating these processes in a more strategic and focused fashion. But it has not put a stop to farmers' inquisitiveness. Farmers have inquiring minds and an ability to find creative, locally adapted solutions. However, they often don't know that they are doing something new and valuable, let alone something that may be applicable in different contexts.

Sharing of knowledge is commonplace

It is a myth to think that farmers do not share ideas. There are, of course, situations in which they will keep knowledge to themselves. One is where sharing will lose them a marketing 'edge'. This may occur, for example, with the selection of a particular variety of crop for its special flavour. Or with a special formula for post-harvest processing. Sharing these ideas will be detrimental to the innovator. But the great majority of innovations, and the knowledge that stems from these, will benefit both the innovator, and others, equally. Take soil fertility management: improved methods will be better for all, on an individual basis but also for shared ecosystem. And, anyway, most innovations on the land cannot be hidden – or protected without expensive patents.

Knowledge exchange takes place through particular pathways

How does knowledge spread? Currently, extension systems work through a variety of means: visits by agents, training courses, publications, radio, messages on mobile phones, the internet and so on. But these extension systems have only been in place for 150 years – alongside Western scientific research. The original pathways were simpler. And in fact they still exist today. The market place is a great medium for exchange of ideas. Markets and livestock auctions – as sociologists will tell us – have always been much more than places to buy and sell. And the other important channel is travel and visits. While these 'classical' routes have been supplemented with modern methods, they have not vanished.

The present: local innovation endures

It is evident in all of the above that local innovation – here specifically referring to agriculture – has not only driven the enterprise forward since its inception 10,000 years ago, but continues today even in the most isolated places or countries in conflict. This offers hope, especially in the marginal areas where 'scientific answers' are few. Here farmers themselves are the great untapped resource. Up to 200 or 300 years ago, particular innovative practitioners were at the forefront of developing improved systems of farming. But the advent of Western science has masked this process, at least in the majority of countries. And, while it may have put local innovation in the shade, local creativity has not been crushed. It carried on – and thrives when given attention and stimulation. Local innovation does not replace existing research and extension (R&E) systems, but it adds value to them and makes them more accountable to the end users.

Characteristics of local innovators

What characterises a local innovator? Or an innovative community? While this book gives us clues to answer the latter, the former is covered in previous publications. And this is not merely an idle question, but yields important clues about how to identify who are innovators (see Chapter 3 for full discussion of identification and selection). Reij and Waters-Bayer (2001) list what farmer innovators have in common – at least those identified under the project, Indigenous Soil and Water Conservation – Phase II (ISWC). These are:

- Most are men.
- Many are strong personalities.
- Most are relatively old and experienced.
- Most of the widely recognised innovators are relatively rich.
- Exposure to other areas stimulated innovation.
- Most are full-time farmers.
- Creativity and formal education are not correlated.
- Innovative farmers tend to develop integrated farming systems.

While the ISWC project does not claim to have based these conclusions on data, there is little to disagree with in this list. It is interesting to compare it with a more rigorous exercise under a related project, Promoting Farmer Innovation (PFI) which recorded data about farmer innovators and their innovations, post-identification (see Chapter 3 for a fuller discussion of PFI and its heritage). There is a carefully phrased caveat in the book of the same name before data from the first 74 innovators were presented (Critchley *et al.* 1999). It states: 'An obvious but important fact must ... be noted: this is an analysis of *the farmer innovators who were identified*, and of course we cannot be certain that our identification process was perfect'. PFI – which was located specifically in dry zones of East Africa – posed specific questions to the 74 farmer innovators and found the following:

- 76 per cent of those identified were men.
- The average age was mid-40s.
- 92 per cent were full-time farmers.
- 17 per cent of these full-time farmers were retired.
- 30 per cent had come up with ideas independent of any outside influence.
- 14 per cent had been influenced by what they had seen elsewhere.
- A very large majority cited cash/food/production as their primary driving force.
- 30 per cent of the innovations were primarily forms of water harvesting.
- 18 per cent of the innovations were primarily forms of soil nutrient management.

The two lists are broadly in agreement. There are some further points made in Critchley's 2007 *Working with Farmer Innovators*, where innovators are noted to be (not surprisingly) 'imaginative and opportunistic'. Another common denominator is that they blend together local resources – and a saying – heard more than once in the drier zones of Africa – is, 'I don't let a drop of water escape'. Wastes are seen as potentially useful inputs, and recycling is a feature of many innovations, thus another common saying: 'everything has a use' (Critchley ibid). Mutunga and Critchley (2001) further note that 'they apparently visualise patterns of integrating resources and intensifying production that escape others'.

There is also a tendency to optimise economic value of available resources by reallocation within the farm or the landscape: thus, where there is abundant fertility in one part of the farm (e.g. a manure heap), this can be spread widely to maximise the returns from the input. Conversely, where there is too little of a resource, say water (as a result of poor rainfall), for it to be productively used, it may be concentrated – in this case through water harvesting. Innovations that achieve multiple goals simultaneously are also not out of the ordinary. Thus a gully may be reclaimed, preventing erosion and capturing rich sediment for cultivation behind the barrier; similarly pernicious weeds may be removed, desiccated and then used for mulching.

Innovation tends to proliferate around the rural homestead, where there is a bounty of resources. Water, fertility, biodiversity, organic and other wastes, as well as labour – and creativity – are all found in plenty here, and form a rich medium for innovation in food production. It is an interesting hypothesis that the origins of urban agriculture could be explained as 'the homegarden transferred to the city'. Characteristically, urban agriculture has been established on the back of local initiative – there were, historically, no extension agents to advise about agriculture in African cities.

Experience with farmer innovation projects in Africa: the foundation for SCI-SLM

Two key projects that focused on individual farmer innovation in Africa laid the basis for SCI-SLM. The first was Promoting Farmer Innovation, which began in

1997 and ran until 2001. It was funded by the Dutch Government, and the UNDP's (United Nations Development Programme's) Office to Combat Desertification and Drought (UNSO) acted as the executing agency. Technical guidance was supplied by Vrije Universiteit's Centre for International Cooperation. Active in Kenya, Tanzania and Uganda, PFI was targeted specifically at the drier areas, prone to drought and desertification (Critchley *et al.* 1999; Mutunga and Critchley 2001).

PFI developed a comprehensive farmer innovation methodology which was used as the basis for SCI-SLM (see Chapter 3). The key was identification of individual farmer innovators. After characterising them, the focus was on selecting the 'best-bet' innovations and using these as learning points for other farmers. Thus a careful process of cross-visits was established; first between the farmer innovators themselves, and then taking other farmers to visit innovators. Throughout the three countries the same methodology was used, and diligent records were taken of the whole process.

At the end of the project, results were summarised in a brief and well-illustrated report (UNDP-UNSO 2001). The most telling are as follows:

- In Kenya, after less than three years, over 4,400 farmers (nearly 60 per cent of them women) had visited at least three farmer innovators each.
- In Tanzania, of the 60 farmer innovators identified, over half were acting as resource persons for NGOs and the government by the end of the project.
- In Uganda, there were at least 600 adopters of innovations among 'ordinary' farmers; and the farmer innovators had taken up more than five new ideas each (on average).
- Overall, in response to the low level of women innovators in the first round of identification (15 out of 74), a gender sensitisation campaign led to an even ratio (27 men and 30 women) in the second round.

The Indigenous Soil and Water Conservation Project (ISWC) had begun in 1992 with the objective of documenting the extent of traditional systems in Africa. ISWC was funded by the Dutch Government, with funds channelled through the Vrije Universiteit in the Netherlands to the participating partners. By 1996 it had published an overview of the topic with case studies from 15 countries across Africa (Reij *et al.* 1996). This was a significant milestone: but the question was, What next? The answer was to spotlight farmer innovators in seven countries across Africa – basically expanding on what PFI had already started. In 1997, ISWC Phase II was established and ran until 2001.

ISWC, however, took a slightly different approach to working with farmer innovators than PFI. Each country formulated a methodology that it found most comfortable to work with, and there was no emphasis on quantitative targets. ISWC was more concerned with establishing 'joint experimentation' between farmer innovators and researchers, and spreading the concept of 'innovativeness' rather than promoting technical innovations themselves. Once again ISWC orientated itself to publication, and its findings are summarised in an edited volume entitled *Farmer Innovation in Africa* (Reij and Waters-Bayer 2001).

PFI and ISWC ended in the early 2000s and, from their pioneering work, there emerged PROLINNOVA. PROLINNOVA (an acronym formed from 'Promoting Local Innovation') was conceived in 1999, when a group of Southern and Northern NGOs (supported by the Global Forum on Agricultural Research – GFAR, the Consultative Group on International Agricultural Research – CGIAR and the French Ministry of Foreign Affairs) met in France to consider how participatory approaches to agricultural research and development (ARD) based on local initiatives could be scaled-up. After inception funding from the International Fund for Agricultural Development (IFAD), the programme was mainly funded by the Netherlands Directorate General for International Cooperation (DGIS) until 2011. Other donors have also supported specific activities (GFAR, the Technical Centre for Agricultural and Rural Cooperation – CTA, the French Ministry of Foreign Affairs, Rockefeller and Ford Foundations, ActionAid, EED Church Development Service, Misereor, the UK's Department for International Development – DFID and the World Bank). PROLINNOVA has been linked as a 'sister' programme to SCI-SLM since the initial conceptualisation of the latter in 2002. This was not just because of their similarities, but also because of the personnel involved who had shared experience in PFI and ISWC.

From the onset, PROLINNOVA positioned itself as an NGO-initiated international learning network aimed at promoting local innovation in the field of agriculture and natural resource management. In practice this means that, in each participating country, a local NGO convenes the major ARD stakeholders from government, research, extension and education, other NGOs and farmer groups. Over the years, PROLINNOVA has developed into a 'Global Partnership Programme', a community of practice, with over 19 multi-stakeholder country platforms in Africa, Asia and Latin America and a wide portfolio of multi-country thematic initiatives (e.g. Local Innovation Support Funds, farmer-led documentation, community resilience – but also initiatives based on the application of participatory innovation development (PID) approaches to gender, HIV/AIDS, climate change and curriculum development/review). Despite the fact that activities of the country platforms differ according to their experience and strengths, the overall network follows a common focus and methodological approach. The focus of PROLINNOVA is on recognising the dynamics of indigenous knowledge and enhancing capacities of farmers to adjust to change – to develop their own site-appropriate systems and institutions of resource management so as to gain food security, sustain their livelihoods and safeguard the environment (www.PROLINNOVA.net). From the methodological point of view, the programme builds on, and scales-up, farmer-led approaches to participatory development that start with finding out how farmers create new and better ways of doing things.

The concept of participatory innovation development is the core of the methodology and it is an expansion (or evolution) of the participatory technology development (PTD) approach. The term, 'PID' emphasises the fact that participatory research, extension and development processes deal not only with technologies but also with organisational innovation and change, including socio-cultural change such as gender roles. At the heart of PID is joint experimentation, in which farmers – together with support agents – investigate possible ways to improve the livelihoods of local people. Often, it is undertaken by farmers together with development agents, without the involvement of formal researchers. It is not possible for formal research to work together with the millions of smallholder farmers in remote, marginal and highly diverse areas throughout the world. In such areas, where blanket solutions are rarely applicable, local experimentation is needed to find a variety of new ways that work and to adapt new ideas to specific local conditions. PID is meant to support and strengthen this local experimentation process, and – in most cases – the farmers' main partners in PID will be development agents in governmental and non-governmental organisations (PROLINNOVA 2009).

PROLINNOVA has been active in networking and dissemination; findings and lessons learnt are presented in a number of publications, among them a book entitled *Innovation Africa – Enriching Farmers' Livelihoods* (Sanginga *et al.* 2009).

Innovative communities

The case for working with farmer innovators has largely been proven. But what about communities? The hypothesis of SCI-SLM was that communities would also prove to be innovative and responsive to stimulation, contributing to the solution of the serious and associated problems of land degradation and poverty in the drylands (UNEP 2009; Di Prima *et al.* 2013). The existence of community initiatives is evidence of situations where people (from small groups to entire villages) have recognised some common local environmental problem and come together, spontaneously, to address them and seek viable and appropriate solutions. Initiatives coming from the communities themselves stand a much better chance of succeeding, spreading and having a more long-lasting impact on the ground than externally imposed solutions. However, as is the case with individual innovators, the innovative capital of local communities has been largely overlooked by extension agents, researchers and decision makers. The lack of recognition and appreciation is often the result of poor knowledge, interest and awareness as well as the absence of a practical methodology to tap into this valuable resource.

In the book entitled *Innovative Communities* (Velasquez *et al.* 2005), two crucial questions are raised: *What does it take for a community to be innovative?* and: *Why are some communities innovative and others not?* The authors' inquiry in the Asia-Pacific region revealed that, despite community innovation being influenced by basic conditions such as time, location and culture, the existence of a conducive environment that enables people or organisations to plan and act innovatively is a precondition. They define it as the 'enabling innovative environment'. The same study also highlighted a number of essential features that characterise the innovation potential of communities:

- shared community recognition of a relatively urgent sustainable development need or challenge;
- established, flexible, local governance structures that can be adapted to accommodate and sustain multi-stakeholder mechanisms and participation;
- effective consensus-building mechanisms;
- active internal and external communication and networking;

- availability of relevant local culture, knowledge and indigenous practices that can combine with new and introduced ideas and technologies to generate innovation; and
- strong local community leadership and a pioneering spirit.

The authors also note that the existence of diversified and conflicting interests can be both the trigger and a threat to innovation. The combination of external forces/factors (e.g. financial or technical assistance from local governments and NGOs; political instability) with internal forces/factors (e.g. community leadership and participation; aversion or resistance to change) can also play a similar role and either undermine or foster the sustainability of the community initiatives.

Concluding remarks

Local innovation can be a means of helping individuals and communities out of poverty, given its potential to increase production, improve food security and promote risk diversification and suitable adaptation to environmental change. However, local innovation is not a 'silver bullet' solution. It is just part of the answer. SCI-SLM, as the pioneer projects that preceded it, has endeavoured to uncover and make the value of local innovation widely known and recognised. Strong on its solid foundations, rooted in theory as well as practice, SCI-SLM entered new territory characterised by the complexity and dynamics of community initiatives dealing with management of common natural resources. While it has taken until now to document SCI-SLM's approach in a form that will spread it widely, there has certainly been a growing trend in international development to work with innovative communities ('demand-led community development' as it has become known) and simultaneously to recognise the power of cross-learning (through 'learning routes' as this has been termed). SCI-SLM does not claim to have pioneered these approaches, but has championed them across Africa and surely has been a positive influence on this emerging methodology.

References

Chambers, R. (1983), *Rural Development. Putting the poor first.* Longman, London.

Critchley, W. (2007), *Working with Farmer Innovators – A practical guide.* CTA, Wageningen, Netherlands.

Critchley, W., Cooke, R., Jallow, T., Lafleur, S., Laman, M., Njoroge, J., Nyagah, V. and Saint-Firmain, E. (1999), *Promoting Farmer Innovation. Harnessing local environmental knowledge in East Africa.* RELMA and UNDP, Nairobi, Kenya.

Di Prima, S., Critchley, W. and Tuijp, W. (2013), *Methodology for Working with Innovative Communities – The SCI-SLM experience.* UNEP-GEF Policy Brief. UNEP, Nairobi, Kenya.

Grenier, L. (1998), *Working with Indigenous Knowledge – A guide for researchers.* International Development Research Centre, Ottawa, Canada.

Millington, A. (1985), Local perceptions of soil erosion hazards and indigenous soil conservation strategies in Sierra Leone, West Africa. Paper presented at the 14th International Soil Conservation Conference, Maracay, Venezuela.

Mutunga, K. and Critchley, W. (2001), *Farmers' Initiatives in Land Husbandry: Promising technologies for the drier areas of East Africa*. Regional Land Management Unit (RELMA), Nairobi, Kenya.

Pretty, J., Guijt, I., Thompson, J. and Scoones, I. (1995), *Participatory Learning and Action: A trainer's guide*. International Institute for Environment and Development, London.

PROLINNOVA, (2009), Local innovation and participatory innovation development. Concept note. ETC Foundation, Leusden, Netherlands.

Pyburn, R. and Woodhill, J. (eds) (2014), *Dynamics of Rural Innovation. A primer for emerging professionals*. L.M. Publishers, Arnhem, Netherlands.

Reij, C. and Waters-Bayer, A. (2001), *Farmer Innovation in Africa –A source of inspiration for agricultural development*. Earthscan, London.

Reij, C., Turner, S. and Kuhlmann, T. (1986), *Soil and Water Conservation in Sub-Saharan Africa. Issues and options*. International Fund for Agricultural Development, Rome.

Reij, C., Scoones, I. and Toulmin, C. (1996), *Sustaining the Soil*. Earthscan, London.

Richards, P., (1985), *Indigenous Agricultural Revolution*. Hutchinson, London.

Sanginga, P., Waters-Bayer, A., Kaaria, S., Njuki, J. and Wettasinha, C. (eds) (2009), *Innovation Africa – Enriching farmers' livelihoods*. Earthscan, London.

UNDP-UNSO, (2001), *Fighting Poverty through Harnessing Local Environmental Knowledge: PFI final report*. UNEP, Nairobi, Kenya.

UNEP, (2009), *Stimulating Community Initiatives in Sustainable Land Management (SCI-SLM)*, Project document. UNEP, Nairobi, Kenya.

Velasquez, J., Yashiro, M., Yoshimura, S. and Ono, I. (2005), *Innovative Communities – People-centred approaches to environmental management in the Asia-Pacific region*. United Nations University Press.

3 SCI-SLM methodology

Origins of the design

Sabina Di Prima and William Critchley

The methodology used under SCI-SLM is crucial to the project in more ways than one. It was carefully designed for impact and effectiveness and based on experience, not simply on hypothesis. Furthermore, its suitability was to be tested under the project. One output of the SCI-SLM project was to be a further refined methodology that could be applied more widely in initiatives of this nature: component 4 of the project was the development of methodology for upscaling and institutionally embedding SLM initiatives. This chapter – closely based on a SCI-SLM Policy Brief[1] - examines the roots of the methodology, explores the reasons why it was structured in its original form and provides general insights on the methodology itself. Chapter 12 then looks at the experience of utilising the methodology and makes suggestions for its improvement.

The 'Promoting Farmer Innovation' model

The methodology used under SCI-SLM is largely based on successful experience under the Promoting Farmer Innovation (PFI) project[2]. PFI's hypothesis was that uncovering and sharing innovations by individual farmers (or 'farmer families') worked in terms of upscaling sustainable land management. PFI functioned effectively – its documented achievements were assessed in several ways – and were impressive (UNDP-UNSO 2001). Key indicators were: *the number of farmers mobilised* (for example, nearly 4,500 farmers visited at least three innovators each in Kenya); *the number of farmers influenced by innovators* (for example, there were at least 600 adopters of innovations in Soroti, Uganda); *the impact on farmer innovators themselves* (for example, again in Soroti, each innovator had undertaken an average of 5.5 new initiatives since being stimulated by project activities). Had PFI entered a second phase, it would have modified its own methodology; however, it was not extended. But the lessons learnt from PFI were used to inform the development of the SCI-SLM methodology. These were as follows:

- In the context of each project, innovation and innovator must be clearly defined.
- Monitoring and evaluation, as well as impact assessments, are crucial feedback mechanisms.
- Research and hard science should be incorporated to strengthen identification, experimentation and interpretation of results.

- Respect intellectual property rights, even when innovations are 'non-marketable'.
- The 'favoured farmer' syndrome is a danger: avoid farmers who are already 'project pets'; and don't lavish too much attention on too few.
- The danger of a male-biased identification process has to be avoided: women's or family innovations may be hidden behind a man.
- No project can be mainstreamed or expanded if it is not cost-effective.

One difference between the two projects, which influenced their respective methodologies, concerns their regional nature. While both cover a number of countries (Kenya, Tanzania and Uganda under PFI; Ghana, Morocco, South Africa and Uganda under SCI-SLM), the approach to the multi-country nature of the projects differed. Under PFI it was a question of testing the effectiveness of the approach in three neighbouring countries simultaneously. However, under SCI-SLM there was, from the start, a strong and explicit element of South-to-South learning. It was perceived that South-to-South learning and cooperation could offer potentially useful opportunities for taking agricultural development forward through partnerships between countries which offer mutually relevant experiences on 'home-grown' responses to land degradation problems. As we will see in Chapters 9 and 11, this became an important element of the overall methodology – and is one of SCI-SLM's specific strengths.

Promoting Farmer Innovation provided a basic and effective methodology at two levels: 'programme development' (vertical scaling-up towards institutionalisation) and 'field activities' (improvement of innovations and horizontal spread). The reason for this distinction was to be able to communicate to those who work face-to-face with farmers what the actual 'on the ground' activities were and avoid confusing them with concepts and activities which concern those who set up and manage such projects. The PFI methodology, tried and tested with individual farmers, was thus modified for use with communities under SCI-SLM. It was considered that SCI-SLM would function effectively with these modifications, but space was deliberately left open for flexibility and improvement. The refinement and further development of the methodology was set as a component and output of the SCI-SLM project. In practice, this required process documentation (the recording of experiences) and a learning-by-doing approach which allowed the methodology to evolve in response to the different situations and specific conditions in the four countries (Ghana, Morocco, South Africa and Uganda).

SCI-SLM methodology: programme development processes

Building on the PFI experience, SCI-SLM adopted the same overall methodological set up and components but with a number of modifications justified by the shift in focus: from individual to collective SLM initiatives.

The 'programme development' component of the PFI methodology was felt to be equally relevant to SCI-SLM at national level. Thus the programme development processes were maintained, but visualised in a clearer manner (Figure 3.1).

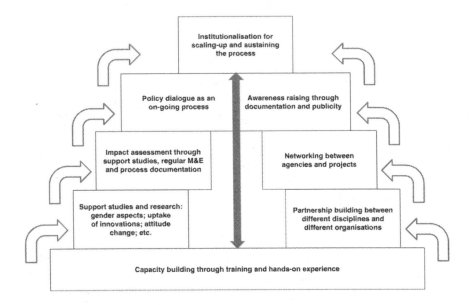

Figure 3.1   Programme development processes used to guide both PFI and SCI-SLM – as visualized under SCI-SLM

Extra emphasis was given to monitoring and evaluation (M&E) and gender aspects through dedicated training sessions and accompanying notes. The programme development component illustrates how a project to promote farmer innovation can be set up and managed, leading to the ultimate goal of integrating this into the existing research and extension system, thereby (a) expanding the source of innovation and (b) improving communication between farmers, scientists and field agents (Critchley 2007).

The programme development processes, which should accompany field-based implementation activities, are represented as building blocks in a pyramid graph, with side arrows symbolising the gradual vertical upscaling. As shown in the figure, the identified processes reinforce each other and pave the way towards the final goal: institutionalisation. Despite the orderly arrangement in the graph, there is no strict sequence in the building blocks of this methodological component – they may start at different moments and run in parallel until the end of the project. However, the training component of capacity building needs to be given high profile, especially at the commencement of the project.

The visual representation of the methodology required inevitably a simplification of a more complex reality. Thus, accompanying guidelines were developed by the project technical advisory group (TAG) to support understanding and interpretation (Box 3.1).

Box 3.1 Programme development processes: summary of guidelines (based on Critchley and Di Prima 2009)

- *Capacity building* is both a foundation stone and essential throughout the project. It comprises training (for example on farmer innovation methodology, participatory innovation development, reorientation of actors' and stakeholders' roles and expectations, etc.) but also learning by doing and reflection. Training and capacity building are not only directed at the core actors, the farmers and the project staff, but also the crucial stakeholders (e.g. community-based organisations, extension agents, researchers, local government officials and policymakers). Capacity building is also a fundamental preface to the introduction of the second methodology component, field activities, which closely depends on effective participation/collaboration between stakeholders historically related through hierarchical links.
- *Inventory of related programmes and initiatives* as a basis for partnership building and networking. This start-up exercise helps on the one hand to avoid conflicts, detrimental competition and unnecessary replication of activities; on the other hand, it enhances synergies with potential partners. Furthermore, the methodology itself – as well as the project approach and the activities – can be improved or sharpened by learning or adapting from other projects or organisations working in the same thematic area and/or location.
- *Partnership building and networking* between relevant agencies/organisations and initiatives. In principle, it is strategic and more sustainable to forge (multi-disciplinary) partnerships founded on mutual benefits; the chance is higher that these partnership alliances will endure beyond the project's lifespan.
- *Support studies* to understand context and address knowledge gaps (for example studies on gender, livelihood, environmental sustainability and socio-economic aspects, spread of innovation and dissemination channels, historical accounts of development interventions in the project sites, etc). Support studies may also result from joint experimentation as well as conventional research. Throughout the project, university students may be instrumental in conducting specific studies to support field work or focus closely on particular issues. Their involvement is mutually beneficial.
- *Impact assessment* to gauge project impact and learn lessons from what worked well and what didn't. The assessment should be carried out at crucial stages of the project and should be based on a combination of quantitative and qualitative data generated by the programme's M&E (see Figure 3.2 on field activities, in particular Steps 3, 4 and 6). It is essential that a farmer innovation initiative can clearly show achievements in quantitative terms to prove its value and cost-effectiveness.
- *Awareness raising* about the concept of local innovation and its potential to relevant audiences. This comprises of a multitude of activities, including targeted reporting, radio broadcasts, videos and briefing notes, as well as articles and posters aimed at spreading the project findings. Awareness creation is

necessary at all levels from local to national and even internationally when project results may have an impact at a higher scale (e.g. global environmental benefits such as carbon sequestration). In any case, it should be based on evidence (data and photographs) and not simply on anecdotal experiences.

- *Policy dialogue* both at national and local levels to lift the project above the local and temporary scale of impact. A change in policy is usually needed to establish an enabling environment in which local innovation can thrive. In this respect, an engaged and influential steering committee at national level will play a crucial role in guiding and influencing relevant policies, based on the documented value/evidence of the project (evidence-based advocacy). Advocates must also demonstrate the connection between field evidence of impact and government priorities (e.g. adaptation to climate change, poverty reduction, food security).
- *Institutionalisation* (integration) of the concept and methodological approach to local innovation within existing (national and local) research and extension systems (vertical scaling-up). The test of factual institutionalisation is whether an approach has become an accepted principle and has developed into common practice.

SCI–SLM methodology: field activities

The 'field activities' component, though maintaining the same ten-step iterative approach of the original PFI methodology, underwent a thorough redesign to capture the numerous aspects and dynamics that intervene when operating with community innovations rather than individual innovations. The original ten-step field activities used under PFI is presented in Figure 3.2.

Why the differences between SCI-SLM and PFI? At the core of this question is the fact that an individual is usually less complex to address than a community. Six issues stand out:

- Defining 'what a true community initiative (CI) in SLM is' can be quite challenging if different disciplinary interpretations are taken into account. What are the key elements that would bring consensus on the working definition?
- The innovation itself will often be of a management nature, rather than technical, when a community is involved; the design of the SCI-SLM project tends to support that logic.
- Characterisation of a group is more complex than simply focusing on an individual. The higher level of complexity depends on numerous factors such as the different roles played by members of the group, status and power relations, the decision-making structures and cost and benefit sharing.
- The concept of ownership of the innovation is also less clear cut when dealing with a community. In fact, within a community it will be natural to have a smaller number of drivers/initiators who take the lead in developing/adapting the initiative and a comparatively larger group of followers.

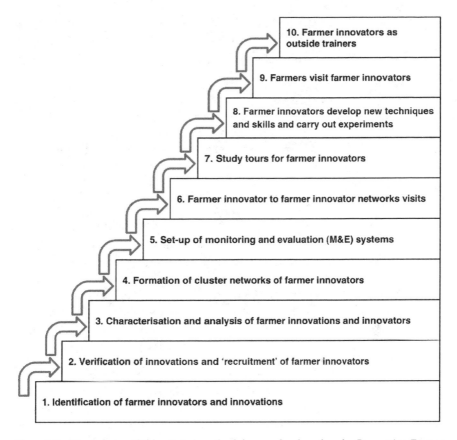

Figure 3.2 The ten-step 'field activity' methodology as developed under Promoting Farmer Innovation (Critchley *et al.* 1999)

- Communication – it is easier for an individual to articulate his/her innovation, than for a community. This is particularly true when an initiative consists of multiple components and the knowledge on the individual components is spread among the community members. Who understands the initiative best in its entirety (holistically)? Who is representative of the group? Who represents the group?
- The cross-visit programme entails careful consideration of who within the group should be involved in the cross-visits. Also, a community can explain its innovation more effectively when demonstration takes place in its own social and natural environment.

In relation to the first point, 'what a true community initiative is', the SCI-SLM team reached a common understanding and agreement on the basic concept by identifying some essential defining criteria – a sort of common denominator that everybody would feel comfortable with. The adoption of a working definition was a paramount decision to actually start implementing the project. For analytical purposes, SCI-SLM's core concept of a 'community initiative in SLM' was deconstructed and the three constituent elements clarified.

A *Community* was defined as a socially coherent group of people (ranging from a small family group, a women's group, a youth group or a water users' group, to an entire village) involved in collective action and having common interest, values and goals in relation to the management of the land and the natural resources in a specific geographical location. The community must be located in an arid or semi-arid region, relying on the available natural resources for its subsistence and livelihood. The group identity should be acknowledged by the community members and should be (formally or informally) recognised by external parties. The boundaries of the community (size, composition, spatial and geographical conditions) are defined by and around the innovation itself.

An *Initiative* was defined as a system (technical and/or social innovation) that is new in local terms; it has potential to spread and generate significant impact (address common problems/needs); and it is developed or adapted by a local community using its members' creativity and with little or no external support (money or other material help). Within the larger group, the innovation process is often led/stimulated by an individual or a small number of members with charisma and leadership qualities. In the SCI-SLM definition, the terms 'initiative' and 'innovation' are used interchangeably. Innovation is a neutral term: just because something is new does not mean that it is necessarily better than existing alternatives. The concept of 'new' refers to an initiative that it is not older than 20–25 years (as a general rule). The initiative should not be an old tradition, a common best practice or an adopted recommendation. A 'technical innovation' can be a product or a process, while 'social innovation' refers to new forms of institutional arrangements – new ways of organising to sustainably manage the natural resources of the community. In more detailed terms, a 'social innovation' can be defined as the process of creating or renewing systems of social order and cooperation which govern the behaviour of a set of individuals within a given human community, with the aim of improving agriculture and the environment and thus strengthening livelihoods. Community initiatives are locally appropriate, by definition, and may include potential lessons of local, as well as global, significance.

Sustainable Land Management is, in simple terms, about 'people looking after the land' – for the present and for the future. The main objective of SLM is thus to integrate people's co-existence with nature over the long term, so that the provisioning, regulating, cultural and supporting services of ecosystems are ensured. In Sub-Saharan Africa, this means that SLM has to focus on increasing productivity of agro-ecosystems while adapting to the socio-economic context, improving resilience to environmental variability, including climate change, and at the same time preventing degradation of natural resources (Liniger *et al.* 2011). The type of community initiatives that the SCI-SLM project actively pursued (based on identified needs) fell under the following SLM categories: water management, protection of trees/forests, community range management and fertility management.

Taking into account these conceptual premises and the highlighted differences, a revised ten-step methodology was drawn up for SCI-SLM, as shown in Figure 3.3. The alterations should be clear in the light of the above points.

The *field activities* 'ladder' guides the process of identification, selection, charaterisation, analysis, improvement, documentation and dissemination of community initiatives through ten steps. The final goals are further stimulation, empowerment, improvement and spread of community initiatives. Ideally, the spread would be twofold: (i) spread of specific (technical and/or social) innovation; and, more importantly, (ii) spread of the concept of community innovation itself.

There is no strict sequence in the field activities but, logically, some steps need to come earlier than others. Despite the numbering of the steps, the methodology remains iterative and dynamic (flexible to change/evolve/adapt based on interpretation, including country-specific factors). This part of the methodology is of particular relevance for those who work face-to-face with communities, namely the fieldworkers and project field staff.

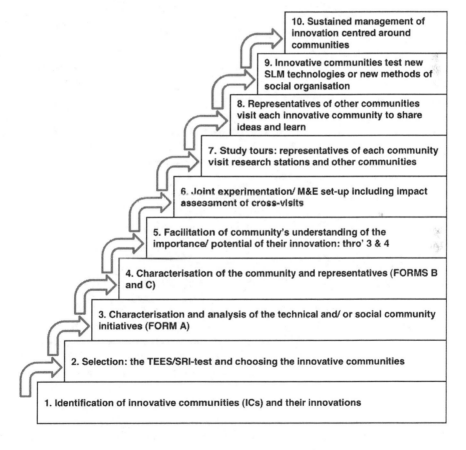

Figure 3.3 Field activity methodology – as designed for Stimulating Community Initiatives in Sustainable Land Management (Critchley and Di Prima, 2009)

As it did for the programme development processes, the visual representation of the methodology for the field activities required, inevitably, a simplification of a more complex reality. The accompanying guidelines developed by the project technical advisory group (TAG) were particularly useful for the training of trainers as part of the capacity-building activities within the project.

Given the relevance and usefulness for other related projects, a detailed outline of the *field activities guidelines* is included in this chapter (based on Critchley and Di Prima, 2009).

Step 1: Identification of innovative communities and their innovations

Through *formal* (e.g. Ministry of Agriculture extension agents, innovation events/competitions) and *informal knowledge networks* (e.g. village or church gatherings, village chief, students, friends, etc.). In simple terms, this step implies looking for evidence of a novel method of social organisation or an innovative SLM practice initiated by a community through a participatory investigation. As often happens, field staff are already aware of a number of potential CIs but they may not have recognised these as 'innovators' before. In this step, it is essential to 'trace an innovation back to its roots' to find the community/group that originally developed it. Identification should not be biased by gender and age aspects or by the fact that the technical innovations are more easily identifiable than the social ones. These types of warnings/suggestions and clarifications about the crucial concepts underlying CIs should be shared with the project's front-line staff and field agents in a training session held before approaching potential innovative communities. This will avoid confusion and the risk of raising expectations on the side of the communities. Identification is not a one-off procedure. It will occur in waves as the project/programme progresses and more initiatives come to light. Often the 'best' innovations are hidden away and only discovered later. Finally, the identification process should fully involve the potential communities, as this is a first step towards an increased awareness and understanding of what they do and their creative capacity.

Step 2: Selection

The selection process should ideally involve all relevant actors: field agents, researchers/subject matter specialists, project front-line staff and last but not least community representatives. In addition to the novelty of the initiative, other aspects/criteria need to be taken into account in order to make an effective selection of the communities to work with, from the larger set of identified communities.

If the innovation is technical, the selection is done through the 'TEES-test'. If the innovation is social, the 'SRI-test' is used. In cases in which a community has developed both a social and a technical innovation, both tests are applied. The correct application of the TEES-test and the SRI-test requires detailed answers. A Yes/No answer or a recommendation for further investigation should be followed by an appropriate explanation or justification. In any case, either a social or a technical innovation should always entail the use of good land management practices/ technologies and environmental protection/conservation as a prerequisite.

The *TEES-test* is used to determine whether a technical innovation has true potential and may lead to long-term/durable benefits. To pass the test, the technical innovation should be technically, economically, environmentally and socially valid. Sustainability is an aspect cutting across these four dimensions. A combination of quantitative and qualitative indicators is recommended to ensure a more comprehensive assessment. However, the list of indicators to be used at the selection stage (and monitored afterwards) should be appropriate (i.e. suited to the specific technical innovation) and adequate (i.e. it should provide sufficient evidence to support a selection/non-selection decision). Table 3.1 lists the attributes of the technical innovation that the TEES-test helps to verify.

Table 3.1 Application of 'TEES-Test'

Technically effective	*Technical appraisal*: Does the innovation work well? Is its performance as good as or better than current alternatives? To answer these questions and make a thorough assessment, the selection team will need to understand and document the design, the functioning modalities and the maintenance requirements associated with the specific technical innovation and make a comparison with the available alternative technologies.	*Examples of indicators/aspects to check*: • yields; • input requirements and their availability in-situ; • basic soil parameters; • number of trees (estimate); • perceived effectiveness of the technique; and • perceived ease of use by community members.
Economically valid	*Cost-benefit analysis*: Do the benefits generated by the innovation outweigh the costs? Is the innovation affordable to the target group? What type/levels of investments are required at different stages of the initiative? What is the time and rate of return on investments? Is the economic performance of the technical innovation as good as or better than current alternatives? To answer these questions and make a thorough assessment, the selection team will need to understand and document the key economic aspects of the innovation and make a comparison with the available alternative technologies.	*Examples of indicators/aspects to check*: • inputs cost; • labour requirements (establishment and maintenance); • income earning attributable to the initiative; • diversification of household income; • savings (labour and/or time); • market demand/market potential; • perceived livelihood improvement; and • perceived increased purchased power.

Table 3.1 Continued.

Environmentally friendly	*Environmental scan*: Does the innovation cause any negative and/or positive environmental impacts? Is off-site pollution or land degradation caused? Is there a gain in biodiversity or a gain in the total system carbon? What are the impacts on specific ecosystem goods and services? Is the environmental performance of the technical innovation as good as, or better than, current alternatives? To answer these questions and make a thorough assessment, the selection team will need to understand and document the environmental impact of the innovation and make a comparison with the available alternative technologies.	*Examples of indicators/aspects to check*: • land use and land status; • water quality and quantity; • carbon stock; • depth of forest floor; • fertilizers and pesticides use; and • abundance, richness and heterogeneity of species.
Socially acceptable	*Socio-cultural assessment*: Is the innovation anti-social towards those who are not members of the community? Has the innovation good potential to spread to others? Does it benefit women and more vulnerable people? To answer these questions and make a thorough assessment, the selection team will need to understand and document the social dynamic around the technical innovation and make a comparison with the available alternative technologies.	*Examples of indicators/aspects to check*: • community rules and regulations; • perceived social benefits (e.g. recognition, reputation, respect, trust, visibility, social status); • perceived livelihood improvements; • increased food security; • acknowledgement of and respect for indigenous knowledge and practices.

   In case of (partial) negative assessment of one or more of these dimensions, the technical innovation does not pass the TEES-test. It will require improvements. Extension agents, researchers and subject-matter specialists play a crucial role in helping the community to improve their own initiative, for example through joint experimentation. However, it will be up to the selection team to decide if a technical innovation which has not passed the TEES-test can still be a suitable candidate for the project or, alternatively, could be solely involved in the dissemination activities at a later stage in the project. The same principles apply to the SRI-test.

   The *SRI-test* is used to determine whether a social innovation has true potential. A good social innovation will be ideally sustainable, replicable and inclusive. The same considerations made about the use of quantitative and qualitative indicators for the TEES-test are valid for the SRI-test. It should be noted that, at the commencement of the SCI-SLM project, the SRI-test was less developed than the TEES-test. The SRI-test helps verify the attributes of the social innovation (see Table 3.2).

Table 3.2 Application of 'SRI-Test'

Sustainable	Can the innovation endure/ continue over time?	*Examples of indicators/aspects to check:* • evidence of the existence of an organised and active community over time (at least six months); • no/little outside assistance including financial support; • financial returns and financial stability; • propensity to plan and invest; • management and record keeping; • M&E system; • links with relevant stakeholders.
Replicable	Has the innovation any potential for spread to/ adoption by other groups? Does it need to be improved before it is spread to/ adopted by other communities?	*Examples of indicators aspects to check:* • organisational structure; • investment needs; • basic external conditions required (enabling environment); • recognition of the value of the innovation by other communities/ outsiders.
Inclusiveness	Is the innovation open exclusively to a selected, privileged group of people or is it open to everybody, including the most vulnerable? Is it a genuine community which benefits its members indiscriminately of their gender, age, race, religion, profession, social status or other characteristics?	*Examples of indicators aspects to check:* • community rules and regulations (e.g. membership requirements); • rights distribution among community members; • level of community members participation; • evidence of democratic practices, transparency and accountability.

In addition to the TEES-test and SRI-test, there are other selection criteria that need to be systematically considered. The community initiative should:

- be within the capacity of 'ordinary' communities. This generally translates into a community initiative being simple, easy and cheap, and requiring minimal specialised skills;
- conform to the project working definition of innovation and fit with the specific field of interest defined for the project (e.g. forest/tree management);
- provide a viable response/solution to changing socio-economic and environmental circumstances (e.g. increased pressure on scarce natural resources due to population growth) or common needs/problems (e.g. nutrient mining causing a progressive decline in land productivity).

Furthermore, the community itself should be suitable (e.g. in terms of innovativeness capacity/potential, communication skills, little/no reliance on external support, etc.), interested in taking part in the project without material incentives and willing to share knowledge/experience and learn from others.

Step 3: Characterisation and analysis of the technical and/or social community initiative (Characterisation Form A)

The collection of data and information that will be included in the characterisation forms (A, B, C) begins in Steps 1 (identification) and 2 (selection) of the methodology. However, the characterisation of the innovation and the community are artificially separated to give adequate visibility to these important steps. It is during Steps 3 and 4 that the relevant information gaps are filled in, the baseline is completed and the data analysed. This sets the basis for the consequent project activities with specific communities, the M&E and the impact assessment carried out in the final stages of the project. Characterisation Forms A, B and C are an integral part of this chapter (see Annex, pages 37–40). These forms, while maintaining the basic questions, can be adapted to suit different projects and contexts.

In Step 3, the field agents and the project front-line staff specifically gather information on the technical and/or social initiative, such as brief description of the initiative; SLM category; date of start; trigger/motivation behind the innovation; source of the innovation; organisational framework; rules and regulations governing the innovative community; investments needed; benefits realised; problems faced and solutions; existing monitoring and documentation; spread of the innovation; and, of course, a detailed report of the TEES-test and/or SRI-test assessment. This information will be recorded in Characterisation Form A and be complemented with relevant photographs of the innovation.

Step 4: Characterisation of the community and its representatives (Characterisation Forms B and C)

In this step, the field agents and the project front-line staff specifically gather information on the community as follows: name of the community; geographical location; type of organisation and its official status; composition of the community (male: female ratio); management structure; benefits of the initiative to the community overall; and links to other communities. They will also collect information on some of the community members/representatives, such as name, role of the person in the initiative, educational level, main occupation and personal benefits achieved through the involvement in the CI. This information will be recorded respectively in Characterisation Forms B and C.

Step 5: Community members are facilitated in understanding the importance and potential of their innovation and helped to be able to explain this to others

This is a crucial step in the methodology. It builds on the previous activities (especially Steps 3 and 4) and creates a solid basis for the following activities. Step 5 implies a gradual progression towards self-awareness of the community:

i)   Discuss, analyse and understand each CI on a case-by-case basis.
ii) Acknowledge its value and potential (positive and constructive feedback).
iii) Help community members conceptualise what they are doing.
iv) Develop a narrative around their case.
v)   Present their case.
vi) Promote their initiative.

Step 5 is closely linked with the next two steps, which are joint experimentation and exchange visits/study tours. In practice, these steps often overlap. Cross-visits between community initiatives involved in the project (see also Step 7) give community members and representatives the opportunity to showcase their initiative and put into practice their renewed self-awareness. In return, the positive/constructive feedback helps consolidate confidence as well as gain trust and reputation among peers. The exchange visits also stimulate mutual learning and creative ideas for (joint) experimentation.

Step 6: Joint experimentation and M&E

As shown in figure 3.3 joint experimentation is at the heart of participatory innovation development. It is a process involving the community, researchers/subject-matter specialists, extension agents and project front line staff. Each actor plays an active role and contributes with knowledge and experience. Joint experimentation is set up to validate, or add value to, innovations, whether technical or social. The validation rationale applies to those initiatives for which there are doubts, or where more investigation is needed for specific aspects of the TEES-test and/or SRI-test. The idea is that an innovation that provisionally passes the above-mentioned tests has to be validated before being spread to others. Joint experimentation can also be carried out to further improve/add value to an initiative which performs better than standard practices. Where the innovation is excellent already and can be directly spread to others, then only monitoring and evaluation is required. The data collected then feed into an impact assessment carried out against the original baseline.

M&E is not confined to Step 6 of the methodology, but runs throughout the project lifespan. Monitoring at its most basic should cover a set of simple parameters concerning the inputs and outputs of the initiative. A combination of quantitative and qualitative indicators is recommended and should match and complement those used for the baseline (see Characterisation Form A covering the TEES-test and/or SRI-test). The list of monitoring indicators should be decided by the community, researchers/subject-matter specialists, field agents and project front-line staff for each initiative on a case-by-case basis (participatory M&E). The decision is not simply on what to measure, but also for what purpose and by whom. The monitoring of some indicators will be entirely in the hands of the community (e.g. parameters that they want to monitor and can measure themselves). Some monitoring (e.g. soil fertility changes, increase in carbon stocks, etc.) may require specialist help. Monitoring of the exchange visits should also be included in the set-up.

It is important that, from the onset, all project staff realise that the baseline, the M&E system and the impact assessment are vital tools to generate evidence that supports claims for success of the project approach and methodology, on the basis of which upscaling can be justified.

Step 7: Exchange visits/study tours

Representatives of the innovative communities are taken outside their location to visit other (innovative) communities, research stations or other relevant project sites. This is a non-monetary incentive as well as a reward for being part of and contributing to the project. Study tours and exchange visits happen at various moments throughout the project, but always after the innovative communities have been selected. The main goal is to create opportunities for sharing ideas and experiences but also to gain exposure to new ideas (technologies and approaches) so at to further stimulate the innovativeness of the selected communities. As mentioned in Step 5, the exchange visits between innovative communities are also an opportunity to showcase initiatives and consolidate confidence as well as to gain trust and reputation among peers.

For practical reasons, cross-visits first occur among innovative communities within the same area (e.g. district). The geographical proximity enhances the occasions for formal (as part of the project) and informal exchanges but also increases the number of community representatives who can be involved in the visit. Under favourable circumstances, cross-visits among neighbouring innovative communities may set the basis for long-lasting bonds and mutual capacity building. Exchange visits can also occur among innovative communities located in different districts, and at national, regional and even international level. In principle, there are no geographical boundaries to the benefits of an exchange visit but, to maximise the benefits, conscientious planning and facilitation are paramount (content/purpose of the visit, transport, travel documents, language translation, etc.). Community representatives will always be able to learn from each other, independently of their language and cultural differences. Being in a new context with new people actually enhances their level of attention and curiosity. In practice, the geographical scope and the number of exchange visits/study tours depend largely on the budget and timespan of the project.

Cross-visits can be bilateral, when the exchange occurs between two communities, or multilateral. In the latter case, a community initiative is visited by representatives of several innovative communities jointly. After the visit, participants will take home memories of the experience but also, as often happens, a selection of samples and planting materials which need to be thoroughly checked to prevent spread of diseases. Pictures, leaflets and any other visual material collected during the visit will be very useful for the participants to share the experience with other community members. Finally, monitoring and documentation of exchange visits/study tours is extremely important, not just to know 'who' went 'where' but to keep a record of the lessons learnt and their impacts.

Step 8: Dissemination

Representatives of other communities (outside the project) visit the innovative communities to learn about their SLM technologies or social arrangements that have been already tested and validated. The best way for them to be convinced is to see and try for themselves. The innovative communities take the lead in the dissemination process. It is proven that locally trusted community representatives help spread ideas/innovations more effectively than extension agents or outsiders. They are generally proud hosts and good communicators. However, field agents maintain a crucial role in the planning, facilitation and coordination of the entire process. Dissemination is a crucial step for the horizontal spread of the selected community initiatives. It starts a grapevine of contacts which potentially lead to the progressive adoption and adaptation of the original initiatives with the ambition of reaching a widespread impact beyond the scope and lifespan of the project itself. Monitoring and documentation of the dissemination process is essential for the impact assessment.

Step 9: Development and testing of new techniques and/or forms of social organisation

Innovative communities are exceptionally responsive to the stimuli and feedback that they get at each step of the methodology. Cross-visits, study tours, joint experiments but also the exchanges with field agents, the researchers/subject matter specialists, project front-line staff and the other community representatives will be major sources of inspiration for the testing of new ideas/concepts/principles and also for the continued upgrading of their innovations. There is no particular stage when the development and testing of new ideas begins and ends. It hardly fits as a step in a logical sequence as it will be a continuous process. Monitoring and evaluation need to keep up with these developments.

Step 10: Sustained management of innovations

Throughout the steps of field activities, the innovative communities will be assisted and supported in strengthening and expanding their SLM initiatives. The ultimate goal is to empower them with knowledge and confidence so that they can sustainably continue on the innovation path even after the end of the project. This will create direct benefit for the selected innovative communities and also stimulate and set standards for other communities. A key sustainability factor will be the ability of the communities to generate a financial return on their innovations. However, other factors will contribute to their endurance and successful upscaling.

Concluding remarks

The SCI-SLM methodology developed at the onset of the project, though deeply rooted in experience, had to prove its suitability on the ground. It was a good point of departure and a comprehensive guide especially for the project front-line staff. It proved to be effective in many ways but also showed certain limitations. Throughout the project, the methodology underwent a number of subtle adjustments, and new ideas for further refinement were gathered. Chapter 11 looks specifically at the experience with the methodology and makes suggestions for its improvement to increase its potential impact and wider application in similar initiatives.

Notes

1   Di Prima *et al.* 2013.
2   Promoting Farmer Innovation (PFI) was a Dutch Government-funded, UNDP-UNSO coordinated project active in Kenya, Tanzania and Uganda from 1997 to 2001. Technical assistance was provided by the Vrije Universiteit Amsterdam's Centre for International Cooperation.

References

Critchley, W. (2007), *Working with Farmer Innovators – A practical guide*. CTA, Wageningen, Netherlands.

Critchley, W., Cooke, R., Jallow, T., Lafleur, S., Laman, M., Njoroge, J., Nyagah, V. and Saint-Firmain, E. (1999), *Promoting Farmer Innovation – Harnessing local environmental knowledge in East Africa*. RELMA, Workshop Report no 2.

Critchley, W. and Mutunga K. (2002), *The Promises and Perils of 'Farmer Innovation'*. SciDev Opinion article: www.scidev.net.

Critchley, W. and Di Prima, S. (2009), *SCI-SLM Methodology Guidelines*. Leaflet prepared by the technical advisory group (TAG) for the project inception meeting in September 2009. Last revision: 13/09/2010.

Di Prima, S., Critchley, W. and Tuijp, W. (2013), *Methodology for Working with Innovative Communities – The SCI-SLM experience*. UNEP-GEF Policy Brief.

Liniger, H., Mekdaschi Studer, R., Hauert, C. and Gurtner, M. (2011), *Sustainable Land Management in Practice – Guidelines and best practices for Sub-Saharan Africa*. TerrAfrica, World Overview of Conservation Approaches and Technologies and Food and Agriculture Organization of the United Nations.

UNDP-UNSO, (2001), *Fighting Poverty Through Harnessing Local Environmental Knowledge: PFI Final report*. Nairobi, Kenya.

Annex

SCI-SLM Summary Baseline Data: Form A
[to be filled in for each Community Initiative]

Characterisation of Technical and/or Social Community Initiative/Innovation

Date: *Team members:*   *Interviewee(s):*

Name of Community/Location . . . .

Composition of Community . . . .

Technical Initiative/Innovation . . . .

and/or

Social Initiative/Innovation . . . .

Eligibility check-list

Genuine community? . . . .

Their own technical SLM innovation and/or social innovation? . . . .

No/little outside assistance? . . . .
[money &/or assistance]

TEES and/or SRI test compliant? . . . .

---

1 Technical SLM innovation
[only if a technical innovation (though may be both social and technical)]

   a. Type: category and brief description . . . .
       . . . .
   b. When was it started? . . . .
   c. What was the trigger for/ motivation behind the innovation? . . . .
   d. Who was the main source of the innovation? . . . .
   e. Is it: A new idea? . . . .
              A modified tradition? . . . .
              An adapted recommendation? . . . .
              Other? . . . .
   f. TEES-test [explain (detail if possible and expand form as necessary to allow comment)]
              Technically effective? . . . .
              Economically valid? . . . .
              Environmentally friendly? . . . .
              Socially acceptable? . . . .
   g. Extra Investments?
              Labour? . . . .
              Cash? . . . .
              Any outside assistance? . . . .

h. Benefits realised?
   Production? ....
   Economic? ....
   Environmental? ....
   Social/cultural? ....
i. Problems faced? ....
j. Solutions? · ....
k. Area under SLM in this community initiative? ....
l. Spread of innovation? [this baseline info ditto] ....
   To how many other communities? ....
   Method of spread? ....
m. Current links with extension/ research/ NGOs? ....
n. Documentation/ monitoring (dates)? ....

2 Social innovation

[only if a social innovation (though may be both social and technical)]

a. Type: category and brief description ....
b. Associated SLM technology? [If it's only social innovation here is where you fill technical SLM details and you can/should expand this section considerably]
c. Is the associated SLM technology an innovation?
d. When was the social innovation started? ....
e. What was the trigger for/motivation behind the social innovation? ....
f. Who was the main source of the social innovation? ....
g. Is it: A new social arrangement? ....
   A modified tradition? ....
   Other? ....
h. SRI-test
   Sustainable? ....
   Replicable? ....
   Inclusive? ....
i. Extra Investments?
   Labour? ....
   Cash? ....
   Any outside assistance? [i.e. other than cash] ....
j. Benefits?
   Social/ cultural? [of the associated SLM measures] ....
   Production? ....
   Economic? ....
   Environmental? ....
k. Problems faced?
l. Spread of initiative?
   To how many other communities? [This is baseline information]
   Method of spread?
m. Current links with extension/ research/ NGOs?
n. Documentation?

SCI-SLM Summary Baseline Data: Form B
[to be filled in for each Community]
Characterisation of Community

Date: *Team members:*

Name of Community/Location . . . .

Technical Initiative/Innovation (refer to form A) . . . .

and/or

Social Initiative/ Innovation (refer to form A) . . . .

3 Details of overall community

o. Type of organisation (village; common interest group etc)
. . . .

p. Official status? (registered etc)

q. Composition of community
    i.   Number . . . .
    ii.   Male/female . . . .
    iii.   Age structure . . . .

r. Management structure? . . . .

s. When was it started? . . . .

t. Was someone local responsible for starting this community organisation? . . . . .

u. Was an outside agency responsible for starting this community organisation? . .

v. Is the community linked to other communities? If so, how? . . . .

w. Benefits of the initiative to the community? [This is actually repetition of A1/ A2]
            Production? . . . .
            Economic? . . . .
            Environmental? . . . .
            Social/cultural? . . . .

x. What problems are faced [organisationally or technically or other]?
. . . .

SCI-SLM Summary Baseline Data: Form C

Characterisation of Community Members

[Interview 1–3 typical members of the community; fill in a form C for each of them]

Date: Team members:

Name of Community/ Location (refer to form A and B) . . . .

4   Details of community member *representative of the community*

y.   Name . . . .
z.   Address . . . .
aa.   Age . . . .
bb.   Male/female . . . .
cc.   Involvement/role in the initiative? . . . .
dd.   Status in family? . . . .
ee.   Status in community/ relative resource ownership?
      . . . .
ff.   When joined the community? . . . .
gg.   Education level? . . . .
hh.   Main occupation? . . . .
ii.   Current benefits of the community initiative to this individual?
             Production? . . . .
             Economic? . . . .
             Land under SLM (ha)? . . . .
             Other? . . . .

4   SCI-SLM

Innovation begins at programme level

Maxwell Mudhara and Mohamed F. Sessay

Background

Stimulating Community Initiatives in Sustainable Land Management aims to identify, improve and upscale local innovation in sustainable land management by communities in the drylands of Africa. The 'commons' are where many of the most serious land degradation problems are found and it is communities that are key to their management. Whereas many observers lament that a free-for-all 'tragedy of the commons' scenario is inevitable where resources are managed by the community, SCI-SLM is based on the premise that there are many examples of community-based innovations in natural resource management and that these deserve to be uncovered and built upon. It argues that initiatives developed by communities themselves are more likely to succeed in ensuring adaptability to local conditions. Of particular importance are those innovations that have been triggered by changing environmental or demographic conditions. In addition, when institutionalised, it is anticipated that government or non-governmental organisations will absorb innovation-centred approaches into their day-to-day operations.

The SCI-SLM project methodology combined the knowledge and abilities of various stakeholders. It was underpinned by partnerships to jointly combat problems related to communities and issues of community-based land management. The basic rationale behind SCI-SLM is thus that spontaneous community initiatives in sustainable land management can be a valuable weapon against land degradation and poverty in dryland areas. The SCI-SLM approach was developed in areas prone to desertification in four countries across Africa, Ghana, Morocco, South Africa and Uganda. Furthermore, the approach sought to facilitate South-to-South learning.

The project set to generate new knowledge through analysing technical and socio-economic aspects of community-initiated SLM and refining ways of stimulating the participatory development and spread of the best of these systems in areas prone to land degradation. Implementation of the approach was accompanied by the development of a methodology to upscale and institutionally embed SCI-SLM approaches at national levels. The focus was on localised, community-driven innovations in SLM. The project was in line with the priorities of other global initiatives such as the United Nations Convention to Combat Desertification, TerrAfrica and the Strategic Investment Programme. The similarity of priorities ensured that common technical, socio-economic and institutional lessons would be widely applicable within Sub-Saharan Africa and inform future directions of the global environmental programmes. This chapter presents the context and challenges underpinning the SCI-SLM project and in common with the global initiatives.

Context and global significance

The countries that participated in the SCI-SLM project were chosen to represent the different geographical regions of Africa (North, South, East and West), with a range of socio-economic and ecological conditions. A World Bank study on the impact of climate change on African cropland shows that African farms are indeed sensitive to climate, and especially temperature, changes. Small-scale, dryland farmers will be especially vulnerable to increases in temperature.

The impact of climate change is unlikely to be uniform across Africa. In Ghana, 30 per cent to 40 per cent of the total land area experiences land degradation (Republic of Ghana 2002). The main problems experienced are deforestation, overgrazing and soil erosion, water pollution, inadequate supplies of potable water, poaching and habitat destruction. Various socio-economic factors account for land degradation in Ghana. One is population pressure. Although the north is not as densely populated as the south of Ghana, the poor quality of land, poverty and hunger have driven people to exploit resources in ways that are detrimental to the environment. Furthermore, ineffective bushfire control systems and general poor soil and water conservation practices in the past have contributed to land degradation in Ghana. Kasanga and Kotey (2001) pointed out that the confusing land tenure and management systems in Ghana, where both state and customary systems work parallel to each other and often contradicting each other, lead to recurring conflicts over land and are one possible cause of limited investments in sustainable land management at local level.

Morocco has diverse climatic conditions dominated by a Mediterranean climate as rainfall occurs within the cool season, while the warm season is dry. The climate is attributed to the latitudinal location, as well as the influence of the Atlantic Ocean and the Mediterranean Sea and that of the Atlas mountain ranges. Rainfall is variable within seasons and between years, with mean annual rainfall ranging from less than 100 mm in the Saharan bioclimate to 1,200 mm in the humid bioclimate. Drought is the most important and dramatic manifestation of the variability. The country has witnessed the longest drought in recent years (1979–1984 and most of the 1990s) causing adverse impact on agricultural production, natural resources and the environment (acceleration of degradation and resource depletion). Morocco has severe problems of land degradation and desertification, mainly due to climatic changes and human activities. Addressing these problems is high on the country's agenda.

Morocco has many rivers, mostly characterised by seasonal torrents, which are used for irrigation and for generating electricity. Agriculture consumes most of the water, i.e. 80 per cent. Water resources are scarce in spite of the efforts in developing dams for irrigation. The agricultural sector is dependent on the climate and the associated risks. According to the World Bank, Morocco makes up for its lack of water by using non-renewable groundwater reserves during years of low rainfall. These reserves are, however, almost exhausted. It is expected that the availability of water per inhabitant will be reduced by half in 2050. Water consumption needs to be significantly reduced according to the availability of resources, which will decrease as a result of climate change.

Morocco is rich in biodiversity and 3 per cent of the forest cover is currently protected. The increase of human population, as well as urbanisation, poor agricultural practices, over-fishing, mass tourism, increase in exotic species and fires, are some of the threats to the country's biodiversity. Poor agricultural practices include farming on steep slopes, cultivating marginal land and poor irrigation practices which contribute to soil erosion, pollution of the water table and soil salinity. Agriculture is key to the living standards of the most deprived Moroccans, where 46 per cent of the population and 70 per cent of the poor live in rural areas.

South Africa is the southernmost country on the African continent. Nearly 91 per cent of South Africa is arid, semi-arid and dry sub-humid, thus falling within the UNCCD category of dry areas (Hoffman *et al.* 1999). Land degradation is an acute problem in Limpopo province and the eastern side of Mpumalanga and KwaZulu-Natal provinces. Sheet and gully erosion are the main modes of land degradation. The average soil degradation index for former homeland areas is nearly three times greater than that of private farmland areas (Department of Environmental Affairs and Tourism – DEAT 2003).

Domestic livestock grazing practices cause loss of vegetation cover and changes in plant species composition. Significant forms of rangeland degradation are bush encroachment and alien plant invasions. The country has experienced accelerated bush encroachment. Deforestation is also a significant form of vegetation degradation in Limpopo, KwaZulu-Natal and the Eastern Cape provinces.

The impact of climate change is already beginning to affect South Africa as evident from the recent increase in droughts and rangeland fires, on the one hand, and erratic rains and flooding on the other hand. This affects agricultural production and is likely to be worse in the future and will escalate the processes of land degradation.

Uganda has an average altitude of 1,300 m above sea level and is characterised by flat-topped hills with gently sloping sides and broad swamp-filled valleys. The climate is tropical but mild due to the high altitude. Temperatures range from about 16° to 29°C, with 1,000 mm or more of rainfall over most of the country. Rainfall has been erratic since the early 1990s in terms of incidence, duration and amount. Droughts and floods are experienced with the associated catastrophic effects. In addition since the 1990s, the seasons have shifted. Given that Uganda's agriculture is heavily rainfall dependent, the erratic swings in season have increased food and water shortages in the country. Death of livestock as a result of lack of water has been common in the cattle corridor, which has forced pastoralists to migrate with their herds. Other constraints include declining soil fertility, poor methods of cultivation, and soil erosion (Ministry of Agriculture, Animal Industry and Fisheries – MAAIF 2000a).

Threats of land degradation

Land degradation and growing poverty are prevailing in all four countries, and local community innovations are emerging to address the challenge. However, these innovations are often overlooked by extension workers, researchers and decision makers.

Failure to recognise local community knowledge and endeavour concerning the environment constitutes missed opportunities to improve ecosystem management. There is a lack of knowledge and awareness of community initiatives, the lack of capacity to draw on them and an associated absence of methodology for extension staff, researchers and others to include communities in design of sustainable land management projects. Initiatives developed by the communities themselves have a much better chance of succeeding and spreading than technical or managerial solutions imposed from outside. In any case, appropriate and cost-effective 'answers' are rarely forthcoming from formal research and extension systems in the drier ecosystems.

Institutional, sectoral and policy context

Ongoing and previously implemented programmes across the project countries were supportive of the aims and goals of SCI-SLM. A look at the institutional setting shows the laws and policies that supported and influenced the implementation of the SCI-SLM project across the participating countries.

The population of Ghana is about 27m., of which 54 per cent is rural. The annual population growth rate is 2.19 per cent, with a density of about 113 inhabitants/km². Ghana is considered a lower middle income country but with about a quarter of its people living on less than $1 a day. Most of the poor people reside in rural areas and specifically in the three dry northern regions. Many Ghanaians continue to depend on natural resources, which leads to significant pressure on the environment. Basic services such as good infrastructure, access to health care, safe drinking water and primary education are often lacking, especially in the rural areas. In 2002, 68 per cent of the rural population had access to improved drinking water.

Ghana's economy has shown relatively strong growth rates in recent years, which has led to some poverty reduction. This economic growth is partly due to Ghana's political stability. The strengthening of the economy is leading to higher employment levels. Nonetheless, unemployment is still widespread, especially outside the major cities.

In Ghana, poverty is characterised by a divide between the north and the south attributed to the harsher climatic conditions and low soil fertility in the north, combined with limited socio-economic developments and lower educational levels. Development initiatives are concentrated in the south of the country. The government has taken steps to reduce the gap through implementation of various development interventions in the northern regions (Republic of Ghana 2002).

A wide range of conventions, programmes, policies and action plans has been initiated to combat poverty and land degradation in Ghana. Ghana is a signatory to the Convention on Biological Diversity (CBD), UNCCD and the Kyoto Climate Change Convention. These conventions, together with Agenda 21 from the Rio Conference on Sustainable Development, provide the international framework in which programmes on land management and natural resources operate. The principal agents for implementing these conventions are the Ministry of Environment, Science and Technology (MEST), the Ministry of Lands and Natural Resources (MLNR) and the Ministry of Food and Agriculture (MOFA).

In 1991, Ghana formulated a National Environmental Action Plan (NEAP) and started implementing it in 1993 under a project called the 'Ghana Environmental Resource Management Project' (GERMP). Under the GERMP, several sub-projects dealt with land degradation and natural resource management.

The Irrigation Company of Upper Region (ICOUR) ran an environmental restoration programme throughout the 1990s, targeting 'erosion control on non-irrigable areas, lake shore protection to reduce the rate of siltation and sedimentation, and encouragement and assistance to smallholders to establish woodlots, practice agroforestry and amenity planting' (Republic of Ghana 1999: 16).

Several other National Action Plans and programmes deal with land degradation. The Agricultural Development Programme (MTADP) of the Ministry of Agriculture, dating from 1990, sought accelerated, ecologically sustainable agricultural modernisation and production. The National Soil Fertility Management Action Plan of 1998, by the same Ministry (which had by then changed its name to the Ministry of Food and Agriculture) had set to restore and maintain soil fertility through sound land management practices for agricultural intensification. In 1994, Ghana established Forest and Wildlife policy, executed by the Ministry of Lands and Natural Resources, to conserve and sustainably develop the nation's forests and wildlife resources. Ghana's National Land Policy promotes use of the nation's land and all its natural resources in accordance with sustainable management principles and to maintain viable ecosystems.

Ghana has a Natural Resources Management Programme coordinated by the Ministry of Lands and Natural Resources. Many of the programme components were directed at the north of Ghana where SCI-SLM was implemented. SCI-SLM cooperated with the Savannah Resources Management Programme (SRMP) and the Northern Savannah Biodiversity Conservation Project (NSBCP). SRMP initiated interventions in 12 pilot communities on integrated watershed management and biodiversity conservation. The GEF-funded NSBCP was designed to focus on biodiversity.

Morocco's population is increasing rapidly, at an annual growth rate of 1.75 per cent. The population is almost 33m., of which 54 per cent live in urban areas. In Morocco, agriculture is the backbone of the economy as it contributes 17 per cent of the gross domestic product (GDP) and employs half the labour force. Most of the agriculture practised is by subsistence farmers under rainfed conditions, but a modernised sector produces food for export. The high incidence of drought spells, combined with a weak safety net system for the vulnerable population, contribute to high levels of rural poverty.

The country's GDP growth rate slowed to 2.1 per cent in 2007 as a result of a drought that severely reduced agricultural output and necessitated wheat imports at rising world prices. This growth in GDP is inadequate to significantly reduce the levels of poverty and unemployment. Poverty, which had declined from 21 per cent to 13 per cent of the population between 1984 and 1991, bounced back to 19 per cent in 1998/1999 and affects 5.3m. people.

The problems of desertification and land degradation are high on the country's priority list. Four zones have been given special attention: the south, the east, the Rif and the centre. Several plans and actions have been implemented to address land degradation. Actions to be undertaken according to the National Action Plan (NAP) are based on the promotion of a participatory approach to deal with the desertification problem. The central strategy of the fight against desertification is to engage people, the administration services and NGOs

In 1995 the Moroccan government ratified the Convention on Climate Change. In addition, Morocco ratified the UNCCD in October 1995, which resulted in National Action Programmes (NAPs). These programmes emphasise the recognition and use of both *communities* and *indigenous environmental knowledge*. The NAP stresses the importance of capacitating communities to organise themselves. Furthermore, it specifies the limitations of technology development through standard sources (research organisations) and highlights the potential role of 'pilot rural initiatives' especially amongst 'youth and women'.

For the success of environmental programmes, there is a need to assure and strengthen the communication between all partners, especially with the 'user' populations and also between the producers and owners of information. In line with this thinking, the government Environment Department started an educational action plan to make people aware of risks facing the environment. The campaign targeted decision makers, elected representatives, the media, women, children (via schools) and rural populations. The campaign encouraged the idea of collective responsibility and was meant to transform individual attitudes and behaviour, in order to transform a part of the Morrocan population to be a positive element in protecting and improving the environment.

In South Africa, more than half of its 51m. people live in poverty and about one-quarter of all households are trapped in chronic poverty. Around 45 per cent of the population live in non-urban areas and, of these, 85 per cent live in the former homelands. Seventy percent of those living in rural areas are poor and three-quarters of all children live in households with an income below the minimum subsistence level, leading to tremendous resource pressure and degradation particularly in rural areas (Integrated Sustainable Rural Development – ISRD 2000).

Whilst many poor people manage to temporarily escape poverty through temporary jobs, remittances or government grants, they fail to do so permanently. The central aspects underlying chronic poverty in the country are historic asset depletion, rural poverty, high inequalities and HIV/AIDS. Poor rural households do not have productive assets (land, skills and access to strategic resources) to become productive farmers. As a result, land-based livelihood strategies are not contributing meaningfully to improvement of livelihoods. As a result, large sections of both rural and urban poor are not able to escape from poverty unless they receive concerted assistance.

Despite the advent of a democratic political dispensation in 1994, millions of South Africans still face unemployment and poverty, particularly in rural communities. Among other challenges, the broadening of participation in the governance of key and strategic resources has allowed access to natural resources by previously disadvantaged sections of the South African population. Where possible, the management of these resources has been and is being integrated into enhancement of rural livelihoods and economic uplifting of local communities, while at the same time conserving biological diversity in a sustainable manner.

Most of the South African government's current expenditure goes towards poverty relief, reconstruction and development. It is not easy to determine how much expenditure goes into land management. However, the Poverty Relief Fund, which was allocated about R1.5bn per year, was spent on poverty alleviation programmes through several line departments. About R500m. per annum was taken up by the Working for Water programme to protect water resources and other programmes in sustainable land management (DEAT 2003).

The government used several measures to address the legacies of the past, particularly poverty. The Growth, Employment and Redistribution (GEAR) programme set out a macroeconomic stabilisation programme for South Africa by providing the economic structure for micro-economic reform. One key policy measure introduced by the South African government was the Integrated Sustainable Rural Development Strategy (ISRDS), the vision of which was to 'attain socially cohesive and stable communities with viable institutions, sustainable economies and universal access to social amenities, able to attract skilled and knowledgeable people, equipped to contribute to their own and the nation's growth and development' (ISRDS 2000:19). ISRDS intended to transform rural South Africa into an economically and socially stable sector that significantly contributed to national gross domestic product. The strategy was supposed to benefit the rural poor and to target women and youth as well as the disabled. The initial Rural Development Strategy for South Africa was formulated within the general framework of the Reconstruction and Development Programme (RDP). The land reform programme and agricultural policy reforms are important components of rural development.

The principal South African instruments for environmental policy are the National Environmental Management Act (NEMA), the Biodiversity Bill and the Protected Areas Bill. NEMA is intended to promote sustainable development and has strong links with sustainable land management. All these instruments play an important role in land management.

South Africa ratified the UNCCD in 1997 and, as party to the convention, the country committed itself to developing and implementing a long-term strategy to address issues relating to desertification. One measure of the commitment was the development of the NAP – a key instrument to combat land degradation and poverty. The vision of NAP is of 'prosperous and healthy South Africans living in an environment restored and maintained through universal improvement in land management to its beautiful landscapes and productive ecosystems that sustain livelihoods and ecosystem services, for the benefit of current and future generations' (DEAT 2003). The NAP vision was to be achieved through the promotion of sustainable land management throughout the country over three years through effective and efficient institutional arrangements at national, provincial, local and community levels; the establishment of effective and efficient partnerships between government departments, the private sector, civil society, overseas development partners and owners and managers of land; and the alleviation of poverty by promoting sustainable livelihoods and enhanced land management.

The NAP cut across many sectors and provided a framework for partnership between government structures, communities, non-governmental organisations and the private sector to work together by providing a knowledge base, capacity and financial resources to address these issues. The necessary institutional arrangements to meet the country's commitments to the UNCCD were in place and the DEAT was identified as the focal point for the implementation of the convention. A multi-stakeholder steering committee guides the role of the focal point for the UNCCD. The committee involves relevant government departments, some provincial departments and NGOs.

The South African Department of Agriculture, Forestry and Fisheries (DAFF), formerly National Department of Agriculture (NDA) is redressing land degradation and sustainable utilisation of natural resources on a national scale (National Land Care Programme document, undated.). DAFF plays a key role in the development of the country and has a vision to increase long-term productivity and ecological sustainability of natural resources. Land degradation and water scarcity are serious issues influencing natural resource sustainability and hence need to be addressed and safeguarded by national government and related agencies. DAFF is key in addressing the international environmental conventions such as the UNCCD, Convention on Biological Diversity (CBD) and the United Nations Framework on Convention for Climate Change (UNFCCC). DAFF has aligned its policies and strategies with government priorities to alleviate poverty and to enhance economic sustainability. Through its national Land Care programme, DAFF is making efforts to relieve poverty by addressing land degradation and achieving sustainable natural resource utilisation.

Land Care is about encouraging and supporting sustainable land use practices and promoting and raising awareness to develop a resource conservation ethic at local community levels. It is based on the concept and practice of community members providing their time and energy to identify, plan and implement community empowerment programmes. Land Care has a thematic approach to its work, and these themes are grouped into either focused investment (watercare, veldcare, soilcare and juniorcare) or small community grants.

The National Development Agency, established through an Act of Parliament, is another policy response employed by South Africa to deal with poverty eradication and its causes. It is a statutory body contributing towards the eradication of poverty through the provision and facilitation of development funding, capacity building, research and policy development. The National Development Agency provides financial support for projects that have a direct impact on improving the asset base of poor communities through support and building capacity to increase competency and efficiency; fundraising; research to establish a credible database of partners with whom to promote relationships; and dialogue and partnerships to influence policy development at all levels of government and society.

In addition, South Africa has a number of Community-Based Natural Resource Management (CBNRM) principles encapsulated in a range of laws, regulations and programmes distributed across various government departments. Several government departments practise CBNRM as an approach or methodology without synergy, and in most cases with duplication of the efforts. Hence there is a lack of coordinated actions as CBNRM is diffused across different government departments. In many instances these departments have limited capacity to implement policy objectives. The government departments that are involved are primarily those that have direct legislative control over natural resources, i.e. the Department of Water Affairs and Forestry (DWAF), DEAT, Department of Water and Sanitation, formerly part of DAFF, Department of Rural Development and Land Reform (DRDLR) and the Department of Cooperative Governance and Traditional Affairs (CoGTA).

In 2001/2002 various stakeholders explored whether or not South Africa needed an official CBNRM policy, leading to the development of a set of guidelines for the implementation of CBNRM in the country. In August 2003, a set of CBNRM guidelines was published and ways to implement these also developed.

Among other things, the CBNRM policy sought to draw up a specific set of inter-departmental guidelines for dealing with historically disadvantaged people who manage resources communally and who have traditional knowledge to contribute to planning and resource management processes. The CBNRM policy implementation guidelines have stimulated the different stakeholders in the CBNRM sectors to implement and adopt these as policy for sustainable natural resources management. SCI-SLM sought to further enhance this process.

Indigenous forest cover and plantations are an important resource in the livelihoods of rural communities. These cover over 2m. ha of South Africa, approximately 1.7 per cent of the country, which contributes significantly to the welfare and values of the country's population. The forest sector contributes R14.6bn to the country's economy or 4.1 per cent of total export earnings. Forests and vast woodlands contribute to rural livelihoods and income generation (DAFF 2003). Involving stakeholders in forest management created the need for joint identification of needs and innovative ways to meet these needs. Such joint arrangements have created the need for an inclusive forestry policy framework, Participatory Forest Management (PFM), developed in 1999. These changes led to the development of a strategy for PFM to ensure participatory management of state-owned forests. The strategy provides operational directions and support for implementation of PFM in all state-owned forests. SCI-SLM also sought to enhance PFM.

Uganda underwent and continues to be in transformation toward economic growth and poverty reduction, which began in the late 1980s. In the 1990s, the country's gross domestic product grew steadily by more than 6 per cent a year from a low rate of 3 per cent in the 1980s, and from 1992 to 2000 the proportion of the population living below the poverty line declined from 56 per cent to 35 per cent. This remarkable turnaround was achieved through policies linked to investments and economic liberalisation. Nevertheless, challenges in the areas of poverty reduction and sustainable development still remain, including environmental degradation.

Uganda has a typical Sub-Saharan dependence on rural livelihoods based on rainfed agriculture. The agricultural sector is the mainstay of Uganda's economy, providing 80 per cent of the population with employment. It is estimated that, between 1970 and 1997, the rural population increased by 90 per cent and the cultivated area rose by 35 per cent. However, the yields of most crops were stagnant or declined. Between 1987 and 1997, the population grew by some 43 per cent, but food production by less than 20 per cent. While reports on agriculture commonly state that Uganda has some of the best soils in Africa, only a quarter of the soils can be classified as being of medium to high potential. Furthermore, the scientific community point to a problem with declining soil fertility, and this is echoed by farmers' concerns. Fallow periods have been reduced as a response to greater demands on the land. Therefore, poverty and land degradation are considered paramount targets of Uganda's development strategies and policies.

Uganda's fundamental development strategy is to eradicate poverty, driven by the Poverty Eradication Action Plan (PEAP) formulated in 1997 and revised in 2000. As a result, all development-related programmes in the country were designed or formulated to respond to the PEAP aspirations. Examples of such plans were the National Environment Action Plan and the Plan for the Modernization of Agriculture (PMA). The NEAP's overall objective was ensuring sustainable natural resource/environment management. The PMA, on the other hand, is a holistic, strategic framework for eradicating poverty through multi-sectoral interventions, enabling people to improve their livelihoods in a sustainable manner. PMA is part of the PEAP and its vision is 'poverty eradication through a profitable, competitive, sustainable and dynamic agricultural and agro-industrial sector'. The mission of PMA is 'eradicating poverty by transforming subsistence agriculture to commercial agriculture' (MAAIF 2000a:vi).

The PMA stated that natural resources had to be used and managed sustainably. The key resources mentioned are land, water, forestry, wetlands and the environment. It also stated that soil and water conservation methods needed to be researched and appropriate technologies demonstrated to farmers (MAAIF 2000a:xi).

Given this situation, in respect of sustainable natural resource utilisation and management, one of the opportunities identified for increasing land utilisation is increasing and sustaining the productivity of the land currently under-utilisation, (MAAIF 2000b). This entails adoption of sustainable soil and water management practices, including the increased use of improved inputs, undertaking measures that reverse land degradation and improving the opportunities for profitable use of small units of land (MAAIF 2000b).

Alongside PMA, the Ugandan government established the National Agricultural Advisory Services (NAADS) by Act of Parliament in 2001. NAADS has an exclusive mandate over extension services in Uganda. The NAADS programme initially covered six districts in 2001/2002, but an aggressive rollout from 2004 to early 2006 brought the total coverage to 47 districts and 344 sub-counties. According to this plan, all districts had to be incorporated into the NAADS Programme by the end of 2008. NAADS aimed to develop a demand-driven, client-oriented and farmer-led agricultural service delivery system particularly targeting the poor and women.

The overriding goal of Uganda government policies such as the PEAP, PMA and the National Agricultural Research Policy (NARP) is the alleviation of poverty and the maintenance of economic growth, thereby directly increasing the ability of the poor to raise their incomes and are aligned to SCI-SLM.

Stakeholder mapping and analysis

The SCI-SLM project had to consider the institutional setting in the various countries in determining how it would be set up. This section discusses the institutional setting in the participating countries and how that influenced the setting up of the project structures.

In Ghana, the University for Development Studies was the SCI-SLM implementing agency. The university was established by the Government of Ghana in 1992. The university's principal objective is to address and find solutions to the socio-economic deprivations and environmental problems which characterise northern Ghana in particular.

Members of communities in the three northernmost provinces (making up the semi-arid and sub-humid part of Ghana) most vulnerable to poverty and land degradation are considered. They are also key to finding suitable solutions to fight these problems. Other stakeholders are the Savannah Resources Management Programme and the Northern Savannah Biodiversity Conservation Project, both mentioned in the law, policy and institutional context.

In Morocco, TARGA-Aide, a non-profit-making interdisciplinary organisation based in Rabat, was the lead implementing agency in Morocco. TARGA-Aide was set up in 1998 to act as an environmental development association, specialising on sustainable human development and natural resource management. TARGA-Aide receives funds from donors to carry out rural development work in specific parts of Morocco. TARGA-Aide operates in rural areas, particularly in mountainous areas with vulnerable ecosystems and oases. Participation, integrated actions and sustainability (by strengthening local capacity) are key focus areas of TARGA-Aide.

Concerning South Africa, the Centre for Environment, Agriculture and Development (CEAD), based at the University of KwaZulu-Natal, was the primary stakeholder, being the overall coordinating agency for SCI-SLM at regional as well as national level. CEAD is a relatively new centre formed by the merging of the Centre for Environment and Development, the Centre for Rural Development Systems (CERDES) and the Farmer Support Group (FSG). FSG was the implementing agency. FSG has placed strong emphasis on meeting the needs of resource-poor farmers, other land users and development practitioners in sustainable agriculture, natural resource management, institutional development and entrepreneurship.

The national government departments involved in the South African component of the project are primarily those that have direct legislative control over natural resources, i.e. the Department of Water and Sanitation, Department of Environmental Affairs (DEA), DAFF, DRDLR, COGTA and the Department of Cooperative Governance and Traditional Affairs and provincial government structures. Other relevant departments at the provincial level, especially in KwaZulu-Natal, are also important stakeholders in SCI-SLM.

The Government of Uganda, represented by the Ministry of Agriculture, Animal Industries and Fisheries was the national coordinating institution in Uganda. MAAIF already had experience in coordinating other projects such as the 'Promoting Farmer Innovation' project from which SCI-SLM has its roots. Other important agencies involved in SCI-SLM were the National Agricultural Research Organisation (NARO), the Ministry of Water, Lands and Environment and its executing agency, the National Environmental Management Authority. Other ministries involved include the Ministry of Land, Housing and Urban Development and the Ministry of Local

Government. Other Ugandan-based NGOs represented in SCI-SLM included Environmental Alert and Participatory Ecological Land Use Management (PELUM) and Makerere University through the Faculty of Agriculture, Faculty of Forestry and Natural Conservation and the Institute of Environment and Natural Resources.

Communities were the main stakeholders in SCI-SLM across all countries as they were the ones who emerged with initiatives.

Linkages with GEF and non-GEF interventions

SCI-SLM sought to reinforce the goals of other SIP operations, especially in those countries pursuing national SLM programmes such as the country SLM investment programme being built up by UNDP and the World Bank in Uganda. It also built synergies with the UNDP Civil Society engagement project through collaboration with the Small Grant Programme in the countries where it operated. The MSP also has a strong linkage to TerrAfrica. TerrAfrica generates and disseminates knowledge in support of SLM upscaling in Africa. SCI-SLM, therefore, served as a testing ground for and source of new knowledge and financing.

The SCI-SLM approach and philosophy are shared by several other initiatives. Amongst these are the World Overview of Conservation Approaches and Technologies (WOCAT), the Southern African CBNRM Research and Networking Programme, the PROLINNOVA programme on promoting local innovation and the Bright Spots Project. The project was consistent with UNEP Medium Term Strategy (MTS) Sub-programme 3: Ecosystem Management, which emphasises UNEP's role in addressing the environmental dimensions of land use management. The project specifically addressed UNEP's commitment to support the implementation of the UNCCD and specific support to Africa with regard to land degradation through the New Partnership for Africa's Development (NEPAD)'s Environment Initiative. The focus on community initiatives in land management complemented that of projects in UNEP/GEF's portfolio in land degradation in Africa under the land degradation focal area (LDFA) as well as under other focal areas where land degradation is a cross-cutting issue. The project established linkages with a wide range of UNEP-implemented projects in West, East and Southern Africa.

During the planning period of SCI-SLM, discussions were held in Nairobi with UNDP's Dryland Development Centre, which was the coordinating agency for the methodological forerunner of SCI-SLM, Promoting Farmer Innovation. UNDP Kenya currently supports a successor to PFI a project that matches PFI with the UN Food and Agriculture Organization (FAO)'s Farmer Field Schools (known as PFI-FFS).

Consistency with national priorities or plans and with GEF policy and objectives

The project was concentrated in sub-humid/semi-arid regions. It thus fell within the areas covered by the National Action Programmes of the UNCCD of each participating country. It emphasised *communities* and also *indigenous environmental knowledge*, two of the keystones of NAPs. Specific NAP links were drawn to the three countries of the four that had finalised their NAPS.

Ghana's NAP states that 'Community participation in all activities to combat desertification is considered critical to the achievement of the desired impact'. The NAP furthermore repeatedly mentioned that local initiatives and local knowledge were essential in the community participation process. In addition, under the TerrAfrica platform for SLM, the country has a Country Strategic Investment Framework (CSIF) that will identify and prioritise areas of investment in the scaling up of SLM in the country. CSIF is a central pillar for aligning donor support and investments in upscalling SLM at country level.

In Morocco, the NAP stressed the importance of the capacity of communities to organise themselves. Furthermore, it specified the limitations of technology development through standard sources (research organisations) and highlighted the potential role of 'pilot rural initiatives' amongst 'youth and women'.

For South Africa, the vision of the NAP was achieved, amongst other ways, by effective and efficient new institutional arrangements in the national, provincial and community spheres The project also complemented the LandCare programme of South Africa, the aim of which was to improve natural resource management at the local level.

The Uganda NAP specifically mentioned 'support to local level community initiatives' and also held up the PFI project as an example of a success. PFI is the inspirational source of the basic concept for SCI-SLM. Uganda also received support under the TerrAfrica platform for SLM and its CSIF identified priority areas for investment in scaling-up SLM.

Conclusion

This chapter has shown that SCI SLM was underpinned by global environmental concerns, which were recognised by the countries participating in the project. Indeed, the project was complementary to programmes that governments were undertaking in their countries.

References

DEAT (Department of Environmental Affairs and Tourism), (2003), *National Action Programme: Combating land degradation to alleviate poverty*.

DWAF (Department of Water Affairs and Forestry), (2003), *Strategy for Participatory Forest Management*.

Hoffman, T., Todd, S., Ntshona, Z. and Turner, S. (1999), *Land Degradation in South Africa*. National Botanical Institute, Cape Town.

ISRDS, (2000), *Integrated Sustainable Rural Development, Strategy Document*. Republic of South Africa.

Kasanga, K. and Kotey, N.A. (2001), *Land Management in Ghana: Building on tradition and modernity*. International Institute for Environment and Development, London.

Ministry of Agriculture, Animal Industry and Fisheries (MAAIF), (2000a), *National Agricultural Advisory Services – NAADS programme*. Master document of the NAADS Task Force and Joint Donor Groups. MAAIF, Entebbe.

Ministry of Agriculture, Animal Industry and Fisheries (MAAIF), (2000b), *Plan for Modernisation of Agriculture: Eradicating poverty in Uganda*. Government Strategy and Operational Framework. MAAIF, Entebbe.

Republic of Ghana, (1999), *National Report to the Third Session of the Conference of Parties to the United Nations Convention to Combat Desertification*. Government of Ghana, Accra.

Republic of Ghana, (2002), *Second National Report to the Conference of Parties to the United Nations Convention to Combat Desertification*. Government of Ghana, Accra.

5 Community innovations in sustainable land management

Lessons from northern Ghana

Saa Dittoh, Conrad A. Weobong, Margaret A. Akuriba and Cuthbert Kaba Nabilse

Introduction

Ghana, with a total land area of 238,535 km² (about 92,000 square miles), has a population of about 27 m (it was 24 m in 2010 according to the 2010 population census). The average national population density is therefore about 113 persons per square kilometre. It is considerably less in rural areas. Land degradation occurs in all parts of Ghana, but it is most severe in the northern savannas mainly due to unsustainable land use practices. The Northern, Upper East and Upper West Regions, which comprise northern Ghana, cover an area of about 97,702 km² (about 41 percent of the land area of Ghana), with about one-quarter of the country's population. The population in northern Ghana is however unevenly distributed. While most parts of the Upper East Region and the western corridor of the Upper West Region have population densities of between 100 and over 200 persons per square kilometre, population densities in several parts of the Upper West Region and Northern Region range from about 10 to about 50 persons per square kilometre. The three regions are situated in the semi-arid part of the country with a relatively short and erratic rainy season of about 5–6 months a year. Annual rainfall usually ranges between 700 and 1,000 mm.

Land continues to be the most important socio-cultural and economic resource in most parts of both rural and urban Ghana and it is particularly so in northern Ghana. Land, its use, its management and its improvements are central to food security and overall agricultural production and development in Ghana. Over 70 percent of Ghana's population are dependent on natural resources for their livelihood (Asenso-Okyere 2008) and land is the chief natural resource. In the case of northern Ghana, the proportion of the population that is dependent on agriculture and its related activities, and thus land, for their existence is estimated to be in excess of 90 per cent. Thus, managing the land in a sustainable manner is critical for livelihoods now and in the future. The people themselves are very conscious of that, and several indigenous land management systems have evolved over time in many communities in all parts of northern Ghana. They have evolved using mainly low external inputs, that is, inputs that are readily available in their vicinities and are not costly to obtain.

Land degradation and desertification are very serious problems in many parts of these regions. It is claimed that, in portions of north-eastern Ghana, desertification and land degradation have rendered soils so humus deficient that they no longer respond to chemical fertilizer application (Environmental Protection Council 1992). This clearly points to a necessity for the intensification of sustainable land management practices in the area. Land degradation has been mainly due to soil mining as a result of use of 'extractive' farming practices, deforestation, overgrazing and bush burning. The problem is very serious in the northern savannas because there is a lot of dry combustible material in the dry season and several farming and cultural practices as well as socio-economic situations exacerbate the problem. Continuous land fragmentation, which is a result of the land tenure system, is one of several examples of these problems.

The Ghana SCI-SLM project thus concentrated in the northern savannas with its premise that 'there are local community innovations succeeding where formal research recommendations have often failed'. Indeed, there have been many indigenous technologies in medicine, nutrition, water use, biodiversity conservation and agriculture which have been found to be very effective. There are several examples where formal scientific research recommendations have failed but indigenous methods have proved successful. Research has, for example, indicated that the people of the Upper East Region have considerable knowledge of the nutritional and medicinal properties of indigenous leafy vegetables (Amisah *et al.* 2002). Also, some indigenous food preparation methods have evolved to enhance their uptake by the body (bioavailability) (Dittoh 2004). In the area of agriculture, indigenous methods of soil conservation and soil improvements have been the secret behind the continued production of crops in the very intensively cultivated areas of several parts of the Northern, Upper East and Upper West Regions of Ghana. It is however not a secret that several interventions by governmental and non-governmental agencies and even international organisations have been largely unsuccessful.

The Ghana SCI-SLM project, through a series of consultations on local SLM initiatives and innovations and visits to several sites to see the initiatives, undertook the characterisation of six community initiatives and selected four as indicated in Table 5.1. Characterisation of the initiatives involved obtaining detailed information of the who, when, why, what and how of the innovations or initiatives from community members through group discussions (see Figure 5.1).

The communities of the various initiatives are Tanchara in the Upper West Region, Kandiga-Atosali in the Upper East Region and Moatani and Zorborgu in the Northern Region. Figure 5.2 is a map of Ghana indicating the selected SCI-SLM intervention sites. Sites were selected taking into consideration balance (Northern, Upper East and Upper West Regions) as well as costs involved in getting to the sites from Tamale. The main innovation in Tanchara and Zorborgu as indicated in Table 5.1 are community forest conservation and non-burning of crop residues on farm lands, while soil fertility management through different composting methods are the innovations in Kandiga-Atosali and Moatani. The composting methods have been 'out-scaled' in the 2011/2012 season into two other communities, Abimpego and Binduri in the Kassena/Nankanna East and West Mamprusi Districts respectively.

Figure 5.1 Characterisation of community initiatives in Tanchara (Upper West Region) and Moatani (Northern Region)

Table 5.1 Community SLM initiatives and their characteristics

Innovation (community initiatives)	Location	Start date	Source (origin) of innovation	Motivation behind innovation	Brief description of community innovation	Initiative typology at baseline	SLM main technology category	Estimated area under SLM at baseline (2010)
Non-burning and controlled cutting of community forest, and non-burning of crop residues	Tanchara, Lawra District, Upper West Region	1970	A community member got the idea from Goziir, a community in the same district	Loss of rafters	Community members formed a committee to protect community forest and to promote non-burning of crop residue to improve fertility of farm lands	Technical and social innovation	Vegetation and soil conservation	Community forest: 12 ha Farm land: 10 ha
On-farm (heap) composting and non-burning of crop residues	Kandiga-Atosali, Kassena-Nankana District, Upper East Region	1995	An elderly community member (Mr Anamlokiya)	Soil infertility	Innovation is a modified traditional method aimed at reducing the tedium of carrying compost from homesteads to farms	Technical innovation	Soil conservation	4.5 ha
Women group solidarity in composting (using pit method)	Moatani, West Mamprusi District, Northern Region	2003	Neighboring community and an agricultural extension agent of an NGO	High cost of inorganic fertilizers	A modified composting and social organization system that harnesses women's labour resources to increase food production	Technical and social innovation	Gender-based soil conservation	4 ha
Community forest and dug-out protection, and non-burning of crop residue on farms	Zorborgu, Tamale Metropolis, Northern Region	Late 1980s and early 1990s	Member of the community (then District Assembly man)	Lack of rafters and decreasing fertility of soils	A community-based system to protect forest, water source and improve soil fertility of farm lands	Technical and social innovation	Vegetation, soil and water conservation	Community forest: 10 ha Farm lands: 8 ha

Source: Field characterization of initiatives 2010.

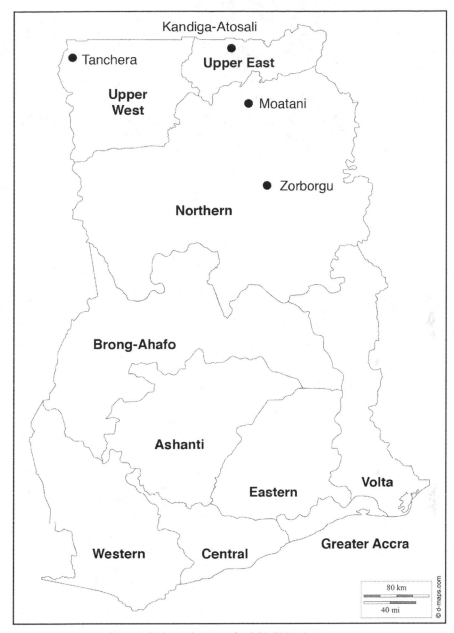

Figure 5.2 Regional map of Ghana showing the SCI-SLM sites

The initiatives and/or innovations by the communities had to meet the TEES and/or SRI conditions. They had to be seen to be:

- **T**echnically effective: being technologically adequate to achieve the objective of the initiative or innovation;

- Economically valid: having the potential to provide some financial benefits;
- Environmentally friendly: being able to improve the environment or at least not to degrade the environment; and
- Socially acceptable: not impinging on community social and cultural values and being acceptable to a large majority of the people. Social innovations in particular also have to meet the SRI conditions.
- That is, they should have good prospects for Sustainability, should be easily Replicable and should be Inclusive in the sense that sections of the community should not be seen to be consciously or unconsciously excluded.

In addition, the following characteristics were taken into consideration in the selection of the initiatives/innovations:

- the number of years the innovation has been practised in the community;
- the number of community members taking part;
- how widespread the practice is, in terms of the number of communities in and around the area practising the innovation;
- evidence of community ownership of the innovation;
- ease of out-scaling and up-scaling; and
- the impact or potential impact on the livelihoods of the people.

It is quite clear that these various criteria and the TEES and SRI conditions are not mutually exclusive.

It can be discerned from Table 5.1 that the main driving force behind all the innovations is the desire of the people to effectively, efficiently and sustainably improve the management of their farm and forest lands in order to improve their livelihoods and to maintain a healthy environment. The inherent soil infertility of the area, the high cost of mineral fertilizers, the continuous destruction of the ecosystem and the soils by bush burning and the decreasing productivity of the land are problems which the indigenous and adapted innovations are aimed at addressing.

The second section of the chapter discusses how Moatani (West Mamprusi) women used social innovation to break tradition with their innovative composting to better their well-being. The third section analyzes socio-economic and other costs and benefits of the various soil improvement methods used in the SCI-SLM sites and how exchange visits between farmers impact upon farmers, probably more than a series of workshops and visits of extension agents. The fourth section of the chapter discusses the revealed disadvantages and benefits of burning and non-burning of forests and crop residues on farms as seen in the SCI-SLM sites. The fifth section discusses the added value of SCI-SLM activities to the community innovations, while the last section draws some conclusions and suggests the way forward with SLM in Ghana.

Moving from LEIA to LEISA through soil fertility management – a challenge to the women of Moatani

Moatani, a small community of about 20 houses, is located in the hinterland of the West Mamprusi District of the Northern Region of Ghana. As at 2012, the estimated population of the community was about 250 people. The main occupation of all adult members of the community is farming; arable crop farming and animal rearing. All households cultivate the same types of crops and raise the same types of livestock. Major crops grown are maize, sorghum, millet, groundnuts, cowpea and soybeans, while livestock raised include cattle, sheep, goats, poultry (chickens) and guinea fowls.

Food crop production in the Mamprusi area of northern Ghana is traditionally a man's responsibility. Men, traditionally, are the breadwinners. Also men are considered the 'stronger' sex and should do the difficult tasks and food crop production in Ghana, generally, is a very difficult and arduous task. Farming in the area is traditionally done with 'no external inputs' or with very 'low external inputs'. Thus the system may be described as either 'low external input agriculture' (LEIA) or even 'no external input agriculture' (NEIA).

The significant agricultural production function for the area may be stated as indicated below:

$$Q = f(A, L) \text{ and not the typical } Q = f(K, L)$$

where Q is agricultural output, A and L are land and labour respectively and K is capital.

Land and labour are the only significant inputs. Capital inputs such as fertilizers, tractor services and others are generally very expensive for the poor farmers or even unavailable at the right time for the few farmers who may afford the services. Even with 50 per cent fertilizer subsidies, most farmers do not use inorganic fertilizers. The use of organic manure on farms is also limited because of its unavailability and the laborious nature of transporting it from homesteads to farms, most of which are located far from communities.

The described situation is definitely worse for women who are regarded as the 'weaker' sex and whose farm labour input is claimed to be limited. Indeed traditionally Mamprusi women and women from several other parts of northern Ghana are not expected to own farms. In any case, land and how it should be used is traditionally in the hands of men. Even how women should use their labour is often guided by men in several communities and households. Many of these situations have, however, changed significantly within the last couple of decades, and many more changes are bound to take place. As stated by one of the women leaders in Moatani, 'these days, if women refuse to farm, there will be no food in the house'. This has been as a result of many changing socio-economic factors. More males go to school and invariably end up in the towns and cities. More males migrate from the rural areas to the urban areas. These and several other reasons have 'forced' women to add more roles in farming activities to their traditional roles of 'home keepers' and 'care-takers'.

Farming activities by women in any part of Ghana and especially in northern Ghana can be very challenging. There is a general issue of the use of productive resources, namely land, labour and manure, etc. The question that is often asked is, 'If men are no longer on the farm as they used to be, what prevents the women from using the resources (e.g. land and manure) the men are no longer using?' Unfortunately the problem is not as simple as that because in addition to the complicated inheritance system, fragmentation due to increasing population pressure and drastic decreases in animals being reared have increased the problems of women's access to land and manure. Women in most parts of Ghana still can only have 'secondary rights' to land. As explained by Hilhorst (2000) 'secondary rights' are of uncertain duration, may not be well defined and may be subject to change with very short notice. Women's rights to land are largely limited to their husbands' rights to land and, in some rare cases, their fathers' rights to land.

A key problem of farming activities in northern Ghana, particularly in the relatively densely populated areas of Upper East Region and the Mamprusi area of the Northern Region is the degraded soils, especially with regard to soil carbon. The situation is even more serious for women because they are usually allocated the most infertile farm lands. Poverty is an additional problem. 'Poverty is a female in rural Ghana' is a statement that is difficult to contest. That means that the women of Moatani are truly practising LEIA, or even NEIA, with resultant deterioration of the soils and the environment over time since they have to supplement their means of livelihood from the exploitation of other available natural resources, particularly available woody species. With the given scenario and other difficulties of transportation and marketing, women in the area are in a vicious cycle of poverty. The breaking of this cycle lies in moving from LEIA to low external input and sustainable agriculture (LEISA). The solution does not lie in moving from LEIA to high external input agriculture (HEIA) as has been attempted by governmental and non-governmental agencies. The natural and socio-economic environmental situations in most parts of northern Ghana cannot support HEIA. According to Dittoh (1999), LEIA is not sustainable, and prevailing economic conditions make HEIA impossible in most parts of northern Ghana. If farmers are left to depend on their present LEIA methods, they are unlikely to survive long in agriculture. Farmers have to improve their present practices and move gradually from LEIA to LEISA by systematically improving their indigenous knowledge of soil fertility management. That is why the story of the Moatani women and their innovative composting and social organisation process is a significant example. Governmental and non-governmental agencies and projects such as SCI-SLM should be able to help communities and groups with such initiatives to move much faster from LEIA to LEISA.

In 1999 one of the women in Moatani visited a neighboring village, Boamasa, and saw a composting method that had been introduced to a women's group in that community by Zasilari Ecological Farms Project (ZEFP), a non-governmental organisation, and thought it was an idea that she could try back in her village. In trying to practise it, however, she found the method too labour-demanding and she decided to involve other women in the community. That is how the Moatani women got together to undertake pit composting. They adapted and improved upon the composting method used at Boamasa by building the pits above the ground and organizing themselves into two formidable groups of women. What is very significant about the groups is the fact that rural women with no formal education in Ghana had the courage to organise themselves to undertake an innovative project to better their lives without recourse to depending on their husbands and the menfolk. This is especially significant as the innovation is based on a traditional activity that is the preserve of men and household heads. A major aspect of the community initiative that is very innovative is the social organisation adopted by the women. The technology was adopted mainly because of the social organisation that went with it. The social organisation allowed the women to pool their energies and ideas together to solve several of the problems that arose from the technology.

The technology involves the construction of a round mud pit about 2 m in diameter and a height of about 1 m, filling the pit with crop residues and other decomposable household refuse, as well as animal droppings, and continuously watering it to hasten decomposition. The larger the animal droppings content of the compost, the richer it is in terms of its fertility potential. The compost is applied mainly to maize farms. The maize crop does not do well without fertilisation. It is improved varieties of maize (*obatampa* and *dodze*). Maize has become an important food and cash crop in almost all parts of Ghana. According to several of the women interviewed, yields on manured (organically fertilized) maize plots are more than four times the yields from unmanured plots in the same vicinity. Whereas 5 bags (of 50 kg each) are often obtained on the average from an acre of unmanured maize plot, as much as 20 bags are often obtained from an acre of manured plots.

As of September 2012, there were two groups of 15 and 16 women, most of whom have their own individual compost pits (Figure 5.3). They do not have group compost pits but help each other to maintain the individual pits. Some of the women are yet to build their own pits either because they are yet to access land to farm or they have other constraints, including their husbands not giving them permission. They also help each other to cart the manure to their farms which are quite some distances from the community. The groups meet regularly to discuss their progress and problems. This has helped them to know of other possibilities in helping themselves. The initiative, apart from its several benefits, has effectively broken certain socio-cultural barriers. Women in the area traditionally do not undertake composting and thus cannot produce certain crops and/or produce at certain levels. The initiative has made it possible for the women to increase production of both a food and cash crop.

The Zasilari Ecological Farms Project, which is a member of the project core team (implementing partner), and the SCI-SLM project assisted the two groups to acquire donkeys and donkey carts to help them convey the compost to their farms (Figure 5.4). By composting and having the means to cart manure to their farms to ensure that the degraded farm lands are resuscitated, the Moatani women are moving from their LEIA or NEIA status to LEISA status. The lands will be improved in a sustainable manner if the composting and farm manuring continues. It is likely that, with increased yields and good marketing arrangements, the women's poverty levels will be reduced. The Ghana SCI-SLM project with Zasilari can assist the women's groups to work along a value-chain to ensure increased sustainability of the process. A possible threat to sustainable production by the women is the possibility of land owners reallocating the present lands being used by the women to other uses. Another problem which the women mentioned is the shortage of materials used for the compost, including maize stalks, soybean vines, groundnut husks and vines, dung, ash and other decomposable materials. As they try to increase the volumes of compost, they will face this problem.

Figure 5.3 Compost pits at (top) Moatani and (bottom) Binduri (an 'out–scaled' site)

Figure 5.4 Donkeys and carts for transportation of compost to farms

Apart from the provision of donkeys to the Moatani women's groups, the Ghana SCI-SLM project encouraged the women in their activities. Several visits by people within and outside the country to the community encouraged them further. The women are also being encouraged to use manure on home gardens to produce vegetables. The out-scaling to Binduri can be attributed to the attention given by SCI-SLM and other groups to the Moatani women's groups. Table 5.2 is a summary historical timeline of the Moatani women's groups composting initiative.

Table 5.2 Historical timeline of the Moatani women's group pit-composting initiative and SCI-SLM intervention*

Date	Activity	By
1999	Observation of composting technology by Boamasa community and introduction of it to Moatani community	Moatani women
1999	Formation of Moatani women's groups	Moatani women's groups (initiated by group and supported by ZEFP, a local NGO)
1999/ 2000	Adapted composting technology by Moatani women's groups (at old community site)**	Moatani women's groups (initiated by group and supported by ZEFP, a local NGO)
11/2009	Scoping visits for promising community SLM initiatives	Ghana SCI-SLM team
06–08/ 2010	Selection and characterization of community initiatives	Ghana SCI-SLM team (with support from partners)
02–03/ 2011	Relocation of community to new site and building of compost pits at new site	Moatani women and their husbands
09/2011	Donation of donkeys by SCI-SLM project to Moatani women	Ghana SCI-SLM team, ZEFP and Moatani community
09/2011	Exchange visits – Moatani women to Kandiga-Atosali	Moatani, Kandiga-Atosali, Ghana SCI-SLM team and partners
04/2012	Exchange visits – Kandiga-Atosali community members to Moatani	Moatani, Kandiga-Atosali, Ghana SCI-SLM team and partners
06/2012	Intercountry visits – Moroccan SCI-SLM team (and farmers) and TAG visit Moatani community	Moatani, Kandiga-Atosali, Morocco and Ghana teams and partners, TAG member
09/2012	Visit of all SCI-SLM teams to Moatani community (as part of Steering Committee meeting)	Four country SCI-SLM teams, UNEP, TAG, all Ghana communities and partners

Source: Ghana SCI-SLM team 2012.

* SCI-SLM project interventions in the other communities were of similar nature.

** The whole community had to relocate to a new place as a result of several environmental, developmental, spiritual and other challenges at the old site.

Comparative socio-economic and environmental analysis of soil improvement methods in three SCI-SLM sites in northern Ghana

The composting method used by Moatani women has been described as being very labour-intensive and the transportation of the heavy manure to farms is also a major constraint. The women of Moatani tried to solve these and other problems by 'group action'. They worked together and sought assistance as active groups and were supported with donkey carts and donkeys. They used the pit method, which involved constructing a mud pit; filling it with trash, household refuse and/or animal droppings; watering it regularly; turning the material in the pit to help it decompose faster; scooping it out; and transporting it to the farms.

The people of Kandiga-Atosali in the Kassena-Nankanna East District of the Upper East Region had a different method of composting, which was started by Mr. Anamlokiya (Figure 5.5) and adopted later by almost all households in the community. It is a heap method. It is a simple method of gathering a farm's crop residue at the beginning of the rainy season into heaps at different locations on the farm and allowing the heaps to decompose during the rainy season. Animal manure and household refuse are also often added to the heap. The decomposed material is then spread on the farm at the beginning of the following cropping season, while new heaps are made in new locations on the farm.

It had been a usual practice by farmers in almost all parts of the Upper East Region, and still is in some communities, to gather crop residues and burn them off in the process of preparing the land for cropping. This was then a novel community innovation, satisfying both the TEES and SRI conditions. The method is technically effective since crop yields have increased as a result. It is also economically valid as financial rewards are expected from the increased yields. The method leads to significant environmental improvements and is socially very acceptable. Over 80 percent of community members practise the method and it is being adopted in other communities. A number of households in Abimpego in the same district have adopted the method (see Figure 5.6) and interactions with the community members indicate its general acceptability by the people. The prospect for sustainability of the practice is also very high as little external effort is required, it is easily replicated and is highly inclusive, in that it involves all members of the society: men, women and the youth.

The heap-composting method of the people of Kandiga-Atosali avoided several of the difficulties faced by the Moatani women. There is no construction of any structure, watering is done by the rain during the rainy season and the manure does not have to be carted to any distant farm as the composting is done on the farm. It must however be noted that the farming systems in the two locations are different to some extent. Even though the concept of compound and bush farms are common to both areas, the major farms of the people of Moatani are 'bush farms', far from homesteads, while the farms considered as major by the people of Kandiga-Atosali are 'compound farms', very close to homesteads. Thus, while in Moatani the compost is carried mainly to bush farms, compost is used mainly in compound farms in Kandiga-Atosali. The main constraints of the Kandiga-Atosali heap method are the very slow rate of decomposition, to the extent that some of

Figure 5.5 (Top) Ghana SCI-SLM team members with the originator of heap composting in Kandiga-Atosali, Mr. Anamlokiya (right). (Bottom) A household head stands next to her compost heap

the crop residue does not decompose until later seasons. In addition, the heaps encourage the breeding of snakes, scorpions and black moles. As the heaps are within compound farms close to homesteads, this poses danger to family members, especially children.

A third soil improvement method has been the non-burning of crop residues on farms by the people of Zorborgu. As explained earlier, burning of crop residue and trash as part of land preparation for cultivation has been a normal practice. Non-burning of crop residue on farms does not involve any pit construction, nor any heaping of the refuse. The crop residue and trash is simply left on the farm to decompose during the following rainy season. This soil improvement method involves even less labour input than the Kandiga-Atosali method.

All three methods are useful in adding organic matter to the soil and thus improving soil fertility and subsequently crop yields. As expected, however, the increase in yields of treated areas are highest in the Moatani case (about 250 per cent), followed by the Kandiga-Atosali case (about 158 per cent), and the least increase is with respect to the Zorborgu case (about 100 per cent), as indicated in Table 5.3. It must be recognised, however, that these relative effects are with respect to a short-run analysis and may or may not be so in the long run. The long-run effects or impact will be substantially determined by improvements made to the practices and their sustainability. Neither improvements on the methods nor their relative prospects for sustainability can be predicted with any certainty. The comparison given in Table 5.3 is only indicative because the analysis focuses on different physical and socio-economic environments and different farming systems including types of crops cultivated and different technologies with widely varying costs. There is thus no concrete basis for objective comparison of sites. What is most important in Table 5.3 is the very significant increases in yields in all cases. The Moatani case is particularly noteworthy even though the cost involved in achieving those increases is far greater than the other cases, as indicated in Table 5.4. The annual cost of producing and applying compost using the Moatani method has been estimated to be about 3.4 times that of the Kandiga-Atosali and 4.4 times that of Zorborgu (Table 5.4). It is important to emphasise the short-run nature of this analysis. That is not, however, to imply that short-term gains are not important or relevant in sustainable land management (SLM). SLM must benefit the people in both the short term and the long term. Practices should, however, not be such as to promote significant trade-offs between short-term and long-term benefits.

Table 5.4 indicates that it is only in the Moatani case that 'capital' investment of 45 Ghana cedis (about US $23) will be required to raise the compost structure. That amount seems meagre; however, to the poor rural woman farmer in Moatani, it is substantial. In addition, in Moatani the women farmers have to either head carry or hire donkey carts to carry the manure to their farms. The provision of donkeys and their carts by ZEFP and the SCI-SLM project significantly improved the timeliness and effectiveness of that process. Timeliness in all farm practices in the area is very critical for productivity because of the erratic nature of rainfall and the need to pay attention to several crop production sites at almost the same time.

Table 5.4 compares the incremental benefits and costs (that is, benefits and costs above practices without the use of the technologies in the various areas) of the three

Figure 5.6 The heap compost method in Kandiga-Atosali at beginning of rainy season and at Abimpego (an 'out-scaled' site)

SLM technologies. It is clear from the table that the Moatani compost pit technology results in a far greater incremental profit than the others.

Table 5.3 Comparison of yield increases due to different soil improvement practices (community SLM innovations)

Community SLM innovation	Crops cultivated	Average farm size (acres) (of the crops of interest only)	Yield (bags/ acre) in farms practising methods	Yields (bags/ acre) in farms not practicsing methods	Percentage increase in yields
Heap composting (in Kandiga-Atosali	Early millet*	2.20	4.0	1.5	166.7
	Sorghum*		5.0	2.0	150.0
Pit Composting (in Moatani)	Maize	1.42	17.5	5.0	250.0
Crop residue management (in Zorborgu)	Maize	4.50	10.0	5.0	100.0

Source: Socio-Economic Survey 2011/2012.
* The two crops are cultivated as mixtures. There are other very minor crops such as leafy vegetables and okra on the plots.

An important agro-economic impact of the three soil improvement methods has been their effect on striga. Striga is a very destructive weed found mainly on cereal crop farms. It is found mainly in very degraded soils and results in very poor growth, and hence yields, of cereal crops such as millet, sorghum and maize. In all the farms practising the three soil improvement methods, the incidence of striga had reduced, according to farmers interviewed. The farmers at Moatani and Kandiga-Atosali noticed that, where the compost is spread, striga is often absent. Some farmers in Kandiga-Atosali even noticed that striga incidence in farms that applied inorganic fertilizers were far higher than farms that applied the organic manure. While there is a definite need to test these several observations experimentally, there is little doubt as far as the farmers are concerned of the efficacy of their compost.

Environmental benefits of the soil improvement methods were also observed by the farmers. The higher yields that result in greater food security and higher incomes to the farmers reduce the desire of community members to cut trees to obtain firewood or charcoal to sell. This point was particularly stressed in Moatani and Zorborgu, where there are nearby savanna forests. The improved soil organic matter that results from all the three methods leads to greater water retention of the soil, less erosion and greater sustainable agricultural production, and many of the farmers have knowledge of that. The Moatani women during their regular meetings educate themselves on these benefits with the help of ZEFP extension personnel, which makes them even more committed to continue the practice and to improve upon it. Organic manure produced from locally available resources which effectively

Table 5.4 Cost comparisons of the different soil fertility improvement innovations (in GH¢)
(All estimated costs relate to organic manure for an acre of farm land)*

Community SLM innovation	Estimated cost of compost structure (2-year life span)	Depreciated cost of compost structure (second column/2)	Estimated cost of gathering crop residue/ animal manure, etc.	Labour cost in watering and turning compost in structure	Estimated cost of carting manure to farms and/or spreading manure on farms	Estimated incremental harvest and post-harvest labour costs (as a result of higher yields) (see Table 5.2)	Total incremental cost in using soil fertility improvement technology
Heap composting (in Kandiga-Atosali)	Nil	Nil	8.00	Nil	6.00	4.00	18.00
Pit composting (in Moatani)	45.00	22.50	8.00	6.00	15.00	10.00	61.50
Crop residue management (in Zorborgu)	Nil	Nil	5.00	Nil	5.00	4.00	14.00

Source: Socio-Economic Survey 2011/2012.

* None of these costs is paid for. Own (household) labour is used. The costs are only indicative, since actual costs vary widely between farmers, at the different locations and even at different times within the year on the basis of environmental and other conditions.

replace inorganic fertilizers is not only economically beneficial to the farmers and the nation, but it also bestows great environmental benefits to the communities and the nation. It has been pointed out that 'many tropical soils are acidic by nature, and mineral (inorganic) fertilizer speeds up the acidification process' (Worner and Krall 2012). It is a fact that maize yields decreased in several African countries in the 1990s even with increases in mineral fertilizer consumption (Scoones and Toulmin 1999). Organic matter provides a buffering capacity in soils, balancing soil acidity, alkalinity and toxicity, and makes plants more resistant to pests and diseases (Hipolito 1999). Thus, the use of organic matter reduces the need for use of pesticides and other inorganic chemicals on farms. It is indeed argued that 'wherever possible organic fertilizer (manure, compost and green manure) should meet the need for basic nutrients, with mineral fertilizers used only to cover any shortfall' in tropical soils (Worner and Krall 2012).

A major encouragement and learning opportunity provided by the Ghana SCI-SLM project was that of exchange visits between Moatani women (and men) and the people of Kandiga-Atosali, as well as a visit by the Moroccan SCI-SLM team to three soil improvement SCI-SLM sites. The exchange visits (although short) had great impact as both Moatani women and men as well as Kandiga-Atosali men and women interacted and learnt from first-hand discussion about the innovations. Both

Table 5.5 Incremental benefits and costs compared

Community SLM Innovation	Crops cultivated	Incremental yield per acre (in bags)	Average price per bag in 2012 (in GH¢)	Incremental benefit at 2012 prices per acre (in GH¢)	Incremental costs (as computed in Table 5.3) (in GH¢)	Incremental profit* (in GH¢)
Heap composting (in Kandiga-Atosali	Early millet	2.5	40.00	100.00	18.00	217.00
	Sorghum	3.0	45.00	135.00		
Pit composting (in Moatani)	Maize	12.5	40.00	500.00	61.50	438.50
Crop residue management (in Zorborgu)	Maize	5.0	40.00	200.00	14.00	186.00

Source: Socio-Economic Survey 2012 and Tables 5.3 and 5.4.
* Per acre profits (benefits less costs) will increase by these amounts because of the adoption of the community innovation.

sides agreed that there were advantages and disadvantages in both composting technologies but wondered if it were not possible to leverage the advantages of each and experiment with a kind of integration of the two methods. Indeed a few women in Moatani have started experimenting with the Kandiga-Atosali heap-composting method. The exchange visits also helped to encourage the men of Moatani to take interest in the work of the women. They had actually not been disinterested but they also did not want to interfere and interrupt what the women were doing. Although they believe that women's groups should stay as women's groups, they did agree that they should assist them in some areas. The men already assist women in building the pits and in the carting of the compost to their farms. The farmers of Kandiga-Atosali learnt from their Moroccan colleagues of the importance of water harvesting so that they could have water in the dry season to water their compost heaps in order to hasten the decomposition of the crop residue.

Burning and non-burning on farms and in forests: social and economic costs and benefits

The non-burning of crop residues on farms was alluded to in the previous section with respect to the Zorborgu community in the Tamale Metropolis in the Northern Region. Non-burning on farms is a practice that is now widely accepted as a very important sustainable land management practice in many parts of the northern regions of Ghana. It is practised in all the four SCI-SLM sites in Ghana. It is, however, unfortunate that burning of farm residues is still practised by some farmers in these same and several other communities. The questions that arise are: Why do some farmers still burn crop residues with all the available knowledge of the benefits of non-burning? Are there benefits in burning crop residues? Does the burning by some farmers have negative effects on those who do not?

Similar questions also need to be asked with regard to burning and non-burning in forests. Savanna forests burn very easily in the dry season and, unless measures are taken to protect them, they will always be 'accidentally' burnt. Protection and non-burning of community forests had been the community initiatives of the peoples of Zorborgu and Tanchara, the fourth Ghana SCI-SLM site. Unfortunately, parts of both the Zorborgu and Tanchara community forests which they had guarded jealously against bush burning were 'accidentally' burnt in 2010 and 2011, respectively. Just as in the case of burning of farm residues, there is a need to find out what benefits accrue to those who burn forests and what are the adverse effects on the people and the communities. With such information, one may be able to look for sustainable ways of addressing the growing menace of bush burning that takes place even in communities that invest energy and resources into trying to prevent it.

Some of the economic benefits of non-burning on farms have been illustrated in Tables 5.3 and 5.5. It is shown in Table 5.3 that that non-burning of crop residues on farms in Zorborgu increases yields by about 100 per cent. Farm profits are also increased substantially (Table 5.5). Other benefits of non-burning, according to the farmers, are reduction of soil erosion, protection of farmlands from the sun and use of less land to obtain required production levels. According to many of the farmers, the main reason why there is a continuation of burning of farm residues by farmers is the benefit related to reduction in labour requirement. When the farm residue is burnt, ploughing of the land becomes much easier. Though a short-term benefit, it is an important one. Many farmers are the aged and women in many communities; thus reduction in labour serves as a very crucial benefit. The inability to plough the land for farming and on time can be a determinant as to whether a farmer can cultivate the land at all in a particular season. Thus the problem does not arise, because the farmers are not aware of the benefits. There is a very strong trade-off between long-term and short-term economic benefits in favour of the short-term benefits. Other reasons given for the burning of crop residue on farms include the existence of snakes and scorpions in non-burnt crop residues, as well as the increase of rodents and insects that destroy crops.

With regard to the community forests, community members in Zorborgu especially were particular about the importance of fruit-bearing trees such as shea (*Vitellaria paradoxa*) and dawadawa (*Parkia biglobosa*), as well as the protection of their dug-out from drying during the dry season. Other major benefits of the forest were the numerous medicinal plants. According to the Chief of Zorborgu, Naa Mahama Yakubu, every plant God created is in their forest and 'there are herbs in the forest to cure every disease'. The community forest is also a source of feed for livestock all year round. The sacred grove is also part of the community forest and it is the duty of every community to protect its sacred grove. They believe that the grove bestows some spiritual benefits on the people and the nation as a whole. There is absolutely no cutting of wood in the sacred grove and it has been so since the coming into being of the community. There is also a dug-out which is the source of water for both humans and animals. The people of Zorborgu believe that conservation of the forest will conserve the dug-out. There has thus been a strong commitment by the community to ensure non-burning in the community forest and even beyond. The forest had been protected by the community for decades. Wild animals started to increase in the forest and, especially at night, they often came onto the Tamale-Yendi road which is close by. Hunters from other communities began to notice the increasing numbers of wild animals and to devise ways to hunt them. 'They succeeded in burning the forest in 2010 to hunt rabbits and grasscutters', lamented Naa Yakubu, Chief of Zorborgu. The community members have been very unhappy that a few grasscutters and rabbits could be the cause of the destruction of such a valuable asset.

The SCI-SLM project started an inventory of the Zorborgu community forest to document the vegetation, the size and composition of plant species and the uses of the species by the people (Figure 5.7). The community forest covers an area of 255,555 m². The most common and the most rare tree species in the forest are as given in Table 5.6. All of them have their economic importance, especially in relation to medicinal properties, and there are herbalists in the community who know what plant is used to treat which disease. The study is on-going and could also be undertaken in Tanchara community forest in the Upper West Region.

SCI-SLM project interventions and out-scaling

The SCI-SLM project in Ghana set out to undertake a number of activities with the sole aim of promoting local SLM initiatives. The activities may be itemised as follows:

- identify community innovations and initiatives in sustainable land management in parts of the most challenging ecosystem areas of Ghana;
- learn from the community members and study the reasons for success;
- identify constraints which they face;
- build capacity in local research and extension personnel to appreciate the important role of local technical and social initiatives;
- raise awareness among policy members with the aim of institutionalizing the community initiatives; and
- stimulate and assist community members to 'up-scale' and 'out-scale' their innovations.

Researchers from the University for Development Studies (UDS) and extension personnel from the Ghana Ministry of Food and Agriculture as well as several NGOs worked together with community members to understand the scientific, economic and social importance of the various initiatives, as detailed in earlier sections. Visits to the various sites by scientists and farmers from within and outside the country gave community members added confidence in their initiatives. There has no doubt been great qualitative benefit to community members as well as researchers and extension personnel in terms of greater belief in community members' ability to be innovative and to solve their own problems. Various constraints were also identified and, through dialogue and discussions, some of these constraints are being tackled. In Moatani, the provision of donkeys and carts by ZEFP and the Ghana SCI-SLM project removed a major constraint. In Kandiga-Atosali, the major constraint tackled was their lack of organisation as a community group to meet regularly to discuss their farm problems and to tackle these as a group. The exchange visits with the

Figure 5.7 Zorborgu community forest, dug-out (during visit of Moroccan SCI-SLM team and SCI-SLM TAG members) and use of measuring wheel to determine size of the forest

Table 5.6 Most common and most rare tree and shrub species in Zorborgu community forest

Most common trees/shrubs	*Most rare trees/shrubs*
Trees	
Acacia gourmaensis	*Daniella olivera* (African copaiba balsani tree)
Afzelia Africana	*Entanda Africana* (Sa-Paanga – local language)
Anogeisius leiocarpus (Sia – local language)	*Ficus spp.*
Azadirachta indica (Neem tree)	*Khaya senegalensis* (Mahogany tree)
Ceiba pentandra (Kapok tree)	*Lannea acida* (Sisibi –local language)
Combretum collium (other Combretum spp.)	*Tectona grandis* (Teak tree)
Diospyros mespiliformis (West African ebony tree)	
Gmelia arborea (White teak tree)	
Parkia biglobosa (Dawadawa tree)	
Piliostigma thonningii	
Terminalia avicenoides (Korli – local language)	
Terminalia moillis	
Vitellaria paradoxa (Shea tree)	
Ziziphus abyssina	
Shrubs	
Annona senegalensis (Wild custard apple)	*Securinega virosa* (Susuwulugu – local language)
Gardenia aqualla	
Rourea coccinea (Eat iron)	

Source: Field Study 2012/2013. (Reference for common names and some local names is Mshana *et al.* 2000).

Moatani community members, the visit of the Moroccan SCI-SLM to the community and the participation of community members in the annual (Steering Committee) meeting of SCI-SLM in Ghana convinced them of the importance of good social organisation in the promotion of farming activities. In Zorborgu, the major constraint has been the need for a well-designed fire belt to protect the community forest from fires that start from other communities. This constraint is in the process of being addressed. In Tanchara, the burning of their forest in 2011 destabilised the community. Their greatest need was therefore to reorganise themselves and also to construct a fire belt. SCI-SLM is encouraging and stimulating the social reorganisation.

Exchange visits by farmers to other farming communities proved to be very beneficial in many respects. Farmers, researchers and extension officers interact quite informally but very effectively during such visits. The intra- and inter-country exchange visits that took place in Ghana provided an important platform for farmer-to-farmer learning and the exchange of valuable knowledge which could never have been transferred by extension personnel or researchers. Moatani is approximately 80 km from Kandiga-Atosali, yet none of the community members in either communities had any knowledge of the technology of the other, even though their basic problem of soil infertility was the same. They only got to know of the different technologies during the exchange visits and this widened their innovative horizons. It is highly conceivable that 'new' innovations can grow out of their widened knowledge of what is possible. Visits to the Zorborgu community forest by the Moroccan SCI-SLM team in June 2012, as well as all the other teams in September 2012, helped the community members resolve to continue their forest conservation activities, despite threats from hunters and other difficulties that they faced in protecting the forest.

Policymakers at the national and NGO levels have been part of the Ghana SCI-SLM team through its steering committee. All the Ghana Steering Committee members are quite influential at national and NGO policy levels and they concur with the ideals of SCI-SLM. This is why institutionalisation of some of the innovations at several levels has been quite smooth. The Ministry of Food and Agriculture in the Kassena/Nankanna East District is promoting the Kandiga-Atosali initiative in other communities in the district, while the ZEPF is intensifying the out-scaling of the pit-composting technology in the West Mamprusi District. For the Zorborgu and Tanchara initiatives, the concentration is on recovery from the bush burnings of 2010 and 2011 and to ensure that better steps are taken to protect the forests. Government establishments and NGOs are ready to assist. These are encouraging developments that might not have taken place without the SCI-SLM project intervention.

Lessons learnt and conclusion

A substantial amount of lessons have been learnt since 2009. The experience of about four years' work with community innovators clearly shows that scientific innovations are not a preserve of so called 'scientists'. Practitioners, such as men and women farmers, are scientists in their own rights and do undertake 'scientific experiments' to arrive at what works and is beneficial for the betterment of their difficult ecosystems. The efficacy of the identified innovations indicates the degree to which farmers themselves have refined their technologies over time. The major value addition to the community initiatives seems to be more in social organisation to help make the systems more economically and socially beneficial than in technical know-how.

Another important lesson learnt is that cultural barriers are not static. Strong social and economic necessities are capable of making societies revise their cultural rigidities. It is gradually becoming clear that men are no longer the 'sole providers' of household food requirements in Moatani and indeed other parts of northern Ghana. The cultural notion of women being mainly home keepers, even in the remotest of societies in northern Ghana, is changing. That means that other cultural and social barriers such as land use arrangements, ownership of livestock and others can be modified if adequate social and economic necessity can be created for their modification.

Another lesson learnt is the need for government and research establishments as well as NGOs to revise their largely top-down research and extension methodologies, especially in the area of agriculture and the environment, and look towards effective incorporation of 'bottom-up' methodologies. Top-down research and extension systems do not pay any attention to very practical and more sustainable local community and individual initiatives.

It was clear from the SCI-SLM project that there are more similarities than differences in the agricultural and environmental problems that confront different communities in African countries. The exchange of experiences between countries across the continent is a vehicle for addressing many of these similar problems. It was very revealing that community members from the Atlas Mountains of Morocco could identify with the problems of community members in northern Ghana during their one-week visit to the region and that an SLM practice in northern Ghana would be of great interest to community members in the Drakensberg region of South Africa.

There has been considerable 'up-scaling' (quality technical and social improvements) in the community initiatives and 'out-scaling' (adoption by others within or outside the communities) as a result of SCI-SLM project intervention. Those activities, once started, will continue as long as the initiatives continue to be socially, economically and environmentally beneficial to the people. The conclusion then is that there are many local community SLM innovations and initiatives in Ghana that have great potential to address several environmental, livelihood and socio-economic problems. There is therefore need to identify these, stimulate their improvement and create enabling environments for their out-scaling and institutionalisation into mainstream governmental and non-governmental programmes.

References

Amisah, S., Jaiswal, J.P., Khalatyan, A., Kiango, S. and Mikava, N. (2002), *Indigenous Leafy Vegetables in the Upper East Region of Ghana: Opportunities and constraints for conservation and commercialization*. International Centre for Development Oriented Research in Agriculture, Wageningen, Netherlands and Centre for Biodiversity Utilization and Development, KNUST, Kumasi, Ghana.

Asenso-Okyere, K. (2008), Living at the expense of future generations: Innovating for sustainable development. Ghana Speaks Lecture/Seminar Series. Institute for Democratic Governance, Accra.

Dittoh, S. (1999), Sustainable soil fertility management: Lessons from Action Research. *LEISA Newsletter* 15(1 and 2), pp. 51–52.

Dittoh, S. (2004), Improving availability of nutritionally adequate and affordable food supplies at community levels in West Africa. In Brouwer, I.D., A Traore, A.S. and Treche, S. (eds) *Food-Based Approaches for a Healthy Nutrition in West Africa. Proceedings of Second International Workshop*. Univ. Ouagadougou/IRD/Wagenningen. University/FAO. University of Ouagadougou Press, pp. 51–62.

Environmental Protection Council, (1992), *Ghana National Environmental Action Plan*. Accra.

Hilhorst, T. (2000), Women's land rights: Current developments in Sub-Saharan Africa. In Toulmin, C. and J. Quan (eds) *Evolving Land Rights, Policy and Tenure in Africa*. DFID/IIED/NRI, London, pp. 181–196.

Hipolito, M.C. (1999), Soil Acidification: Myth or reality? *LEISA Newsletter* 15 (1 and 2), pp. 24–25.

Mshana, N.R., Abbiw, D.K., Addae-Mensah, I., Adjanohoun, E., Ahyi, M.R.A., Ekpera, J.A., Enow-Orock, E.G., Gbile, Z.O., Noamasi, G.K., Odei, M.A., Odunlami, H., Oteng Yeboah, A.A., Sarpong, K. and Tachie, A.A. (2000), *Traditional Medicine and Pharmacopoeia; Contribution to the revision of ethnobotanical and floristic studies in Ghana.* Scientific, Technical and Research Commission of the Organization of African Unity (OAU/STRC), Accra.

Scoones, I. and Toulmin, C. (1999), *Policies for Soil Fertility Management in Africa.* Institute of Development Studies, Sussex and International Institute for Environment and Development (IIED), London.

Worner, B. and Krall, S. (2012), *What is Sustainable Agriculture?* GIZ, German Federal Ministry for Economic Cooperation and Development (BMZ).

6 Community initiatives for sustainable natural resource management in the High Atlas, Morocco

Mohamed Mahdi, Zakariaa Tijani, Mohamed Tamim and Wendelien Tuijp[1]

Introduction: the High Atlas

Morocco has an arid climate which affects 93 per cent of its territory. Aridity, caused by limited rainfall and worsened by land degradation, has increased pressures on croplands, wood and fodder. It is estimated that the forest decreases by 31,000 ha every year, and bush fires damage another 3,000 ha per year. It is again estimated that 8.3 million ha of rangeland (40 per cent of the total area) are severely degraded.

The Atlas Mountains comprise a range across the north-western stretch of Africa extending approximately 2,500 km through Morocco, Algeria and Tunisia. The High Atlas forms the catchment for a number of river systems. The majority of the year-round rivers flow to the north, providing the basis for the settlements there. A

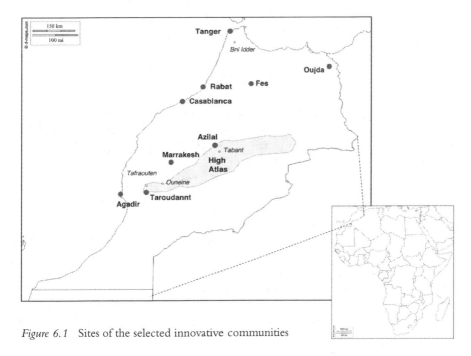

Figure 6.1 Sites of the selected innovative communities

number of seasonal rivers terminate in the deserts to the south and the plateau to the east of the mountains. The Atlas Mountains are rich in natural resources. They contain deposits of iron ore, lead ore, copper, silver, mercury, rock salt, phosphate, marble, anthracite and coal among other resources.

The Atlas serves as a weather system barrier in Morocco running east–west and separating the Sahara's climatic influences, which are particularly pronounced in the summer, from the more Mediterranean climate to the north, resulting in dramatic changes in temperature across the range. In the higher elevations, snow falls regularly, which lasts well into late spring in the High Atlas, mostly on the northern faces of the range. Rainfall varies from year to year between 300 and 900 mm, alternating years of drought and rainfall (Hanich *et al.* 2008).

Location of the communities

The communities that were identified as having promising SLM initiatives and thus selected for the SCI-SLM project are situated in the High Atlas (Figure 6.1). The rural town of Ouneine is located in the valley of the same name. It is connected to the outside world through three tracks: to the north is the winding road that passes through Wijjadane (2,089 m) to reach the national road no. 501. In the south, two forest tracks head for L'khmiss Sidi Ouaziz and Iguer n'Tznart to reach the main road no. 32, linking Agadir and Ouarzazate. The territory of Ouneine valley covers about 260 km².

The rural community of Tabant is situated in the valley of Aït Bougamaz, which can be reached by two tarmac roads leading to both Aït Mhammed town and the city of Azilal. The territory of Tabant town extends over an area of 488 km². It is bordered to the north by the town of Aït Mhammed, to the east by the towns Aït Bou Oulli and Aït Abbas, in the West by the town Zaouiat Ahansal and in the south by the province of Ouarzazate.

Demography

According to the census of 2004, the population in Ouneine was 8,417 people, spread over 1,379 households in 65 villages (or Douars). In Tabant, there are 13,012 people spread over 1,898 households in 25 Douars. The average size of a household is estimated at 6.1 persons in Ouneine and 6.8 in Tabant. Table 6.1 shows that the population in Tabant is higher than that of Ouneine. Furthermore, the average annual growth rate (TAAM) of Tabant is higher (1.2 per cent) than that of Ouneine (0.3 per cent).

Looking at the distribution groups of the population by age of the two towns (Figure 6.2), it is clear that there is a predominance of the working-age group, 15–59 years, which exceeds half of the total population.

Furthermore, in Ouneine, females constitute 53 per cent of the population. This has mainly been because of the migration of many young men to the cities. In Tabant, females constitute 50 percent of the population. The illiteracy rate is high in both areas; 80 per cent of women have not had any formal education. In both areas, however, the people have access to basic social services. Ouneine has municipal

Table 6.1 Demographics of Ouneine and Tabant

Sites	RGPH 1994		RGPH 2004		Rate of growth (%)
	Population	Households	Population	Households	
Ouneine	8,168	1,188	8,417	1,379	0.3
Tabant	11,598	1,663	13,012	1,898	1.2

Source: General Census of Population and Housing 1994 and 2004.

clinics and Tabant has a health centre. Also electricity coverage is quite high, covering about 67 per cent of households in both locations.

It is important to realise that, in Morocco, the average annual population growth rates between 1994 and 2004 were about 0.6 per cent for the rural population, 2.1 per cent for the urban population and 1.4 per cent for the total population. Thus the annual population growth rate in Tabant was far higher than the annual growth rate of the rural population, while the annual growth rate of Ouniene is well below the national and the rural annual growth rates.

One might be tempted to explain these differences by high birth rates in Tabant and a migratory movement out of Ouneine. Table 6.2 gives information on fertility and migration in the two areas. The fertility rate, according to the General Census of Population and Housing 2004, is 3.7 in Ouneine and 4.6 in Tabant. With regard to migration, out-migration is more than in-migration in both communes as shown in Table 6.2.

The 'incoming' are usually newly married women from neighbouring rural communities who join their husbands' families.

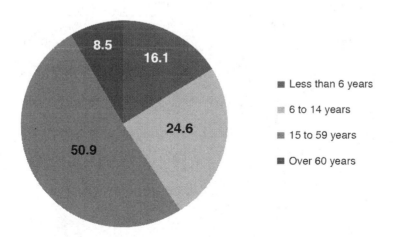

Figure 6.2 Distribution of the population of Ouneine and Tabant by age groups

Table 6.2 Total fertility and migration

Sites	Total fertility*	Migration**
Tabant	4.6%	580 incoming – 2,650 outbound = –2,070
Ouniene	3.7%	1,030 incoming – 1,610 outbound = –580

Source: ★ General Census of Population 2004. ★★ General Census of Population 1994.

The rural town of Tabant is economically richer than that of Ouneine. Tabant has more agricultural land, more water and richer pastures. Tourism also plays an important economic role in the valley of Aït Bouguemaz to which Tabant belongs, with support of the state.

Farming households

The Ouneine and Tabant communities are characterised by small family farms. According to the General Census of Agriculture (1996),[2] there were 757 farms in Ouneine and 1,294 in Tabant (Table 6.3). Several households can cultivate a farm together. Thus the percentage of farms (RGA 1996) compared to households (General Census of Population and Housing 1994) is 64 per cent in Ouneine and 68 per cent in Tabant. This is so because the male heads of households in some young families who have no property rights to land migrate to work elsewhere.

Table 6.3 Ownership structure of Ouneine and Tabant

Sites	Farms	UAA* (ha)	Plots	UAA irrigated	UAA non-irrigated
Ouneine (High Atlas)	757	980	2,635	576	404
Tabant (High Atlas)	1,294	4,238	13,316	3,052	1,185

Source: General Census of Agriculture 1996, Ministry of Agriculture and Marine Fisheries.
★ UAA: Useful agricultural area.

According to the General Census of Agriculture (1996), 757 farms in Ouneine share 980 ha with an average of 1.2 ha per farm, while in Tabant 4,238 ha are being cultivated with an average of 3.3 ha per farm. Farming conditions in Tabant are more favourable: 72 per cent of the cultivated land in Tabant is irrigated, whereas the irrigated land in Ouneine is 41 per cent. A general conclusion is that the communities are characterised by small fragmented farm sizes. Fragmentation is expressed by the number of farm plots, while the small size relates to the size of arable plots.

The production system

The production system in the two sites can be described as agro-pastoralism. In both cases, the system closely integrates subsistence cropping based on cereals with small ruminants and cattle. Grain is produced for food and fodder, and other cereals and legumes for livestock. Livestock are an important source of manure. In Ouneine fruit trees such as carob, almond, olive and walnut are also important in the production systems and are a major source of farm cash income. In Tabant, the introduction of apple trees has boosted the local economy over the past 20 years.

In these mountainous areas, women are the cornerstone of the production system, even though their economic contribution does not seem to be recognised. They are responsible for many tasks within the agro-pastoralism system. Taking care of the cattle is their domain and it allows them to provide their families with milk and butter, and the selling of one or more calves each year contributes substantially to the family income. They are the ones who supply the family with wood and water for fire and cooking. These tasks are both hard and time consuming. The women carry bundles of wood weighing about 30 to 40 kg over several kilometers. The supply of tap drinking water and the arrival of butane gas relieved their burden, but their agricultural and pastoral responsibilities remain and become heavier when the men migrate.

In areas with fragile ecosystems, such as the High Atlas, land degradation has exacerbated the imbalance between a growing population and stagnating or decreasing natural resources. This causes a 'saturation of the environment', in which the population is larger than the resources necessary to support their livelihoods. Faced with this dilemma and the need to find ways to survive, people in communities can either leave their land or come up with innovative solutions to alleviate the effects of environmental degradation. This chapter highlights initiatives by four different communities in the High Atlas of Morocco who have attempted to combat land degradation and to develop innovative solutions. The four initiatives demonstrate the ability of these communities to provide local answers to the problems of natural resource degradation.

Box 6.1 Tribal social structure

Both Ouneine and Tabant are located in the High Atlas mountain range and populated by Berber communities. The Berber identity is usually wider than language and ethnicity and encompasses the entire history and geography of North Africa. The unifying forces for the Berber people could be their Berber language, belonging to the Berber homeland or a collective identification with the Berber heritage and history. In Morocco, after the constitutional reforms of 2011, Berber has become an official language and is now taught, in principle, as a compulsory language in all schools regardless of the area or the ethnicity.

The Arabic language is also spoken by a large segment of the population, particularly by those who have been to school and/or migrated. The ethnic, social structure of the two sites is like a tribal pyramid and is segmental. On top of the pyramid is the tribe of Aït Ouneine and Aït Bougamez (Tabant), which is divided into several segments (first level of segmentation). The segments are subdivided into Douars (second level of segmentation). Today, the tribe has lost its salience in political and social life. The territorial level (Douar or village) is the most functional. It is the territorial unit and the social, economic and religious basis of social organisation. Each is composed of Douar lineages, that is to say, families considered to be descended from the same ancestor. Today the communal division coincides more or less with the contours of the old politico-tribal organisation in the two localities. In addition to blood ties, many economic interests link the families of a Douar, particularly for the management of communal natural resources (water, forests and rangelands). This has created a spirit of solidarity and developed a sense of collective action and initiative among the people.

Methodology

Identification/selection process of community initiatives

The process of site selection under SCI-SLM, and the identification and characterisation of the community initiatives is a process that has gone through several stages. In the beginning, the focus was on two rural towns, Bni Idder in northern Morocco (located approximately 27 km south-west of Tetouan) and Ouneine in the High Atlas. The Targa-Aide research team has been working in these two locations for several decades, together with its national and international partners (in the early 1980s in Ouniene and the late 1990s for the north). Targa-Aide is an independent, non-profit association, created in 1998; its mission is to contribute to the establishment of sustainable development aimed at the reduction of socio-economic inequalities (see: www.targapaide.org).

Pre-identification and discussing SCI-SLM concepts and methodology

In May 2010, a Targa-Aide team (including Steering Committee members), together with members from the technical advisory group, visited rural communes in northern Morocco. The objectives of this visit were to achieve coherence of mutual perceptions on the concept of community innovation; to develop a common understanding of the concepts and methodology of the project; and pre-identification of community initiatives. Though the innovations pre-identified at this stage were interesting, they could not be included in the project because they could not be clearly marked as community initiatives, but rather individual ones. The main result of the visit was a better understanding of the objectives and methodology of the project. The exploratory team was convinced of the need to develop a sensible methodology for identifying community innovations that were standardised and usable by different members of the team and, moreover, valid for the selected sites where project activities would be up and running. It was thus decided to select Ouneine and Tabant as the project sites.

In Tabant, researchers from Targa-Aide have been involved in development projects since the start of the socio-economic dynamics of this region. They are in direct contact with local actors. The population of the region is known for its tradition of social management of natural resources, especially of forest *Agdals*. The question under consideration was whether the dynamic development could lead to collective initiatives for land management.

Applied methodology

The applied methodology developed by Targa-Aide combined two survey tools:

1 interviews with informants to gather information from development agents, farmers and other local stakeholders on innovations of which they were aware of;
2 studies of the agrarian system and the production systems in communities in both sites to identify what was common practice, making it easier to spot innovations. A semi-structured guide was developed to systematically and comprehensively investigate the agrarian system and the production systems of all relevant Douars. The guide explores the practices common to a community and provides data on crops, livestock, social management of natural resources and social organisation and focuses on changes and innovations introduced in these production systems and social aspects. Based on the guide, the team carried out an investigation on the agrarian system and the production system. The investigation allowed the team to identify numerous innovations and four were selected.

This methodology was combined with the tools provided by SCI-SLM. To be selected, all potential innovations had to pass several tests, including the eligibility test, TEES-test and SRI-test, and then had to be characterised according to SCI-SLM methodology. The selection was based on certain criteria: the mix of social and technical innovations; potential for spread; diversification of the portfolio (for example, community initiatives related to forest management, water, crops, etc.); and involvement of partners. As a result, the following initiatives were selected: forest management in Anzi and Agouti communities; land rehabilitation in Machal community; and new rules for allocating water rights in Afourigh community (Figure 6.3, Table 6.4).

Case studies: forest management by two different communities

Anzi community in Ouneine

The community of Anzi, in the Rural Commune of Ouneine, consists of 52 households. Families live on subsistence agriculture based on cereals (barley and maize) combined with livestock. On average, a family owns one cow, eight goats and three sheep. Some families have been practising carpentry for several generations, and their practice is dependent on wood from the forest, owned by the State. The carpenters produce a rich variety of decorative, as well as functional, items. They use ingenious home-made tools to create their artistic products. These include domestic equipment such as tables and stools, tillage equipment such as ploughs,

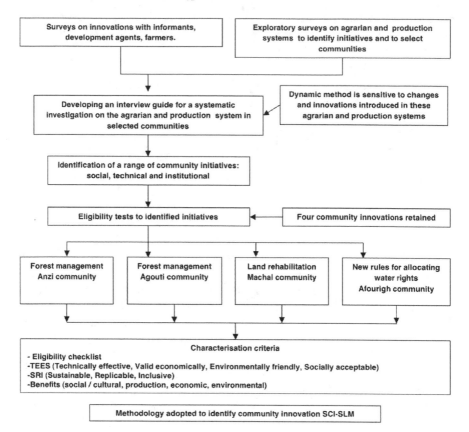

Figure 6.3 Workflow of selection process of community initiatives

handles, shovels, picks and hoes, but also building materials such as doors, windows and frames. The carpenters meet local demands. The number of carpenters has increased over time, and they have become well known in the community. In 2012 the number of carpentry workshops reached 18, including a collective workshop. Subsequently, the carpenters of Anzi focused on manufacturing products for bazaars and decorating town halls, hotels and Riads. Items are treated with chemicals to give them a whiff of antiquity but are not sold as old objects.

Agouti community in Tabant

The Agouti community in the Rural Commune Tabant consists of 80 households. Families depend on horticulture, mainly apple trees that were introduced in the 1990s, cattle and tourism. The latter sector directly employs 16 official guides, three cooks and many related jobs. The community manages two cottages. Tourism has enabled them to develop the tradition of wood crafts that has diversified both the activities of the community and the sources of income. Small handicrafts such as

Table 6.4 Overview of selected communitiy initiatives (CIs) in Morocco

CI characterization	Anzi	Agouti	Lamhalt	Afourigh
Name of CI	Carpenter co-operative potentially leading to reduced pressure on the forest	Cooperative working on handicrafts with part of shared profit invested in tree planting	Land rehabilitation	Water rights, innovative irrigation management system
Location	Ouneine, High Atlas	Ait Bouguemaz, High Atlas	Ouneine, High Atlas	Ouneine, High Atlas
Starting date	2006	2006	1990	1994
Origin/source of CI	Core group	Core group of youth	Individual	Core group of youth
Motivation/ trigger of initiative	To help in marketing of their products and in purchase of timber from the market and to prevent over-exploitation of the forest	Conviction about the value of woodcraft and the awakening awareness about environmental degradation	The inability to cultivate arable land in the direct neighbourhood of their village	A new generation of charismatic leaders who are better educated and willing to realise sustainable land degradation management, and the construction of a new storage basin
CI brief description	Cooperative, based on artisanal woodwork, aiming to formalise their activities. They sell their products both at local and national markets. Before this, only for local use	Members produce small wooden objects for the tourist market. The price is divided into three parts: 1/3 for the producer, 1/3 for the association, and 1/3 invested in tree planting: both fruit trees that are donated to people and trees for communal areas	Three families reclaimed a hillside and converted it to productive irrigated terraces. The improvement of the land included increased crop diversification	Collective management of water rights: a group of young men decided to change the private property rights during periods of water scarcity. The second aspect consists of a change of the irrigation water division structure
Initiative typology	Social	Social and technical	Technical	Social
SLM main technology category	Community forest management	Community forest management	Land rehabilitation	Water management

Source: Targa-Aide, 2010.

spoons, bowls and pen holders are produced to sell to tourists who visit the Aït Bougamaz valley. Carpenters collect dead wood from the communal forest or buy wood (pruned from trees) from land owners. Carpenters work in a collective workshop, equipped with all necessary tools (Figure 6.4). In addition to selling directly to tourists, products are also marketed through a website.

How to manage the forest sustainably – comparing the two initiatives

In Anzi, about 30 local carpenters formed a cooperative in 2006 for various reasons. The origin of their carpentry tradition lies with their grandfathers and before, and they produce a wide variety of products. The main reason for forming the cooperative was to help in the marketing of products and in the purchase of timber from the market.

Carpenters also harvested indigenous wood such as Thuya (*Tetraclinisarticulata*) and green oak (*Quercus ilex*) without legal authorisation. This exposed artisans to fines and criminal convictions. Branches were also cut to feed livestock. Interestingly, the idea of creating a cooperative was inspired by a forest officer, a year before its establishment.

The cooperative provided a legal basis to market their products more easily outside the Douar. The question can therefore be asked as to what extent this initiative is a genuine SCI-SLM initiative. From the social point of view, there certainly is evidence of a group working together under the banner of their cooperative, which is new in this area. From the SLM perspective, carpenters are aware of the need to preserve the mountain forests where the Thuya tree grows. Thuya, which is indigenous to North Africa, cannot be harvested legally from the forest, and carpenters know they can be prosecuted if they harvest the tree. To make this a true SCI-SLM initiative, they had either to come to an arrangement with the Forestry Department to harvest the forest sustainably, or they would have to plant and nurture all the species that they need in their work. The Moroccan SCI-SLM team suggested that a cross-visit to another group of carpenters, who harvest wood sustainably in Aït Bougamaz valley, might trigger such an arrangement.

The relationship between the community and the forest and the success of the SCI-SLM initiative had been due to multiple factors acting simultaneously:

- The supply of wood tended to be limited. For three years, the carpenters had to buy wood in Marrakech, Oulad Berhil, Aoulouz and Taroudant, thus reducing their dependence on wood from the surrounding forest by 50 per cent.
- The extra income through carpentry changed people's life styles and had a positive impact on their attitudes towards the forest. As a result of the extra money, the people increasingly started to use gas for cooking, and currently half the family households cook on gas, which is subsidised. This has reduced the need to cut firewood considerably.
- Although progress has been made since the involvement of the community in the SCI-SLM project, there is still a lot of work to be done. Half of the families still use wood from the forest for domestic fuel, and the community uses wood to heat the five *hammams* (public baths) of the Douar.

Figure 6.4 Community workshop of Anzi

- Since 2009 there has been widespread cultivation of alfalfa (*Medicago sativa*), with the rehabilitation of irrigation canals (about 4,300 metres of canal) and the development of the basin. Alfalfa provides additional fodder resources and eases the pressure on forest pastures by 80 per cent.

In Agouti, 11 young men founded the local Ighrem Association in 2006. 'Ighrem' in Berber means 'collective loft'. Initially, the association was created to meet the basic needs of development and health services. The initiators of the association have received training from the Ministry of Health and a regional association based in Azilal.

Carpentry was considered to be a useful extra income source next to agriculture and tourism. But carpentry was not the sole reason; some of the founders were, and still are, also active in tourism. They were aware of the negative influence of tourism on the environment and they wanted to improve the environment as well. The association's activities concentrate on handicraft, ecotourism and the environment. The association is composed of 34 members including the founding core that consists of very dynamic and active young people. Many of the activities complement each other and are related to environmental protection (tree planting), sanitation and solid waste management. The association is closely connected to the agricultural association *Tamzrite* (Country). There is a special branch for women with about 18 members. The women produce small bracelets and other small handicrafts out of wool, in many colours, traditionally painted with plants and mainly sold to tourists.

With regard to craftsmanship, community innovation lies in the concept of marketing. The association has established a unique rule for revenue sharing of its products: one-third of the revenue is for the crafts, one-third goes to the association and one-third is for environmental protection. The last third is a kind of environmental tax: the association invests one-third of the price for each wooden item sold to buy apple trees and donates them to members of the Douar who are economically and/or socially vulnerable. In August 2009, the association planted 80 apple trees to benefit disabled persons and widows with children (Figure 6.5). In 2010, 108 young trees were distributed to farmers in need. In 2012, the association changed this rule and is now limiting its support to the activities of women in the Douar.

The association also expanded its horizon by starting partnerships with an association called Agharass (Road) consisting of Moroccan and French partners, and with the Peace Corps. These partnerships are helping the association to strengthen its carpentry equipment and its marketing. Initally, the products of the workshop were sold on site but Agharass helped to sell products made by Ighrem carpenters at a fair organised by the American School in Rabat. The association is now active in e-commerce as well; wooden objects can be ordered through their website.

The two communities of Anzi and Agouti share the following characteristics:

- a strong endogenous institutional commitment resulting in mobilisation for collective action;
- a generation of active young people, some of whom have received primary or secondary education and returned to their families. These young people are increasing their living space and interconnectivity, integrating mobility between the village, the city and also abroad;
- innovations consisting of the organisation of craftsmen in a cooperative or association to market their products, which can be categorised as an institutional type;
- community innovations occurring in the context of natural resource management and the fight against rural poverty;
- a level of dependence on the forest strongly linked to the level of development of activities and income;
- women involved in related activities to the initiatives. However, in Anzi women have only started to join the cooperative from 2011 onwards.

Main differences between Anzi and Agouti:

- Anzi is located in Ouneine, which is an isolated valley of the High Atlas, whereas Ighrem (in Agouti community) operates in Aït Bougamaz, which is a valley visited by tourists and reached by tarmac.
- Carpenters of Anzi descend from a long tradition of wood carvers, producing large handicrafts such as doors, tables and ceilings. In Agouti, the carpenters of Ighrem started in the 1980s to produce small handicrafts, such as spoons, bowls and pens, suitable for tourists.
- Anzi carpenters initially produced for the local market only, but this has now changed. Agouti carpenters produce for the international tourist market, including selling via the internet.

Figure 6.5 Evolution of rehabilitated land in Ouneine valley: (top) Construction of terraces; (bottom) Planting trees along the terraces and irrigated agriculture

- In Anzi the selling of handicraft is the main source of income for the carpenters; agriculture is the secondary source. In Agouti, members have multiple sources of income because of the diversification of activities, which complement and reinforce each other.
- In order to get the necessary wood, carpenters in Anzi either illegally harvest wood from the forest or purchase wood from the market. In Agouti members either collect dead wood in the forest, an arrangement with the Forest and Water Department F&W Department), or the wood is bought directly from tree owners after the pruning of their trees.
- At the beginning of the project, environmental awareness in Agouti was higher than in Anzi. This was one of the reasons for the start of the community initiative.

Emergence of environmental awareness in the two communities

In the context of advanced degradation of the environment, it is feared that the development of wood crafts dangerously affects forest resources. This assumption justifies the project support to the people of both communities in their efforts to develop activities that sustainably manage forest resources, upon which they depend.

Both the Anzi and Agouti communities are trying in their own ways to implement original forms of organisation and are merging their woodcraft with sustainable management of forest resources.

In Aït Bougamaz (where Agouti is located), there has been deforestation on communal lands, as can be seen when looking at the surrounding hillsides. The Department of Water and Forestry (DWF) has done some replanting, mainly pine trees, which is not highly appreciated by the population. Most local people would like to have types of bushes and trees that can provide fodder, whereas pine does not fulfil this purpose. That is why most people do not want the replanting by the Department to be close to the village. The Ighrem Association has an agreement with the Department, which stipulates that twice a year members can harvest a determined quantity of dead wood from the communal forest. The types of wood to be collected are mainly juniper and box wood. Carpenters buy walnut from the owners of the land on which the trees grow.

The quantities of wood used for objects are relatively insignificant considering the type of objects produced. In the past people used to divide the communal land/forest into several parts. This system is known as Agdal (see Box 6.2). As Ighrem members have become more and more aware of environmental issues, they have developed new ideas to protect the environment. Together with the Council of the Douar, called *Jmaa 'at*, it is decided which area is to be kept fallow for a certain period, but problems arise with neighbouring villages that do not respect the practice. In order to protect and develop the forest, Ighrem members are in contact with the Forest Service about the establishment of a tree nursery with forest species and finding ways that lead to social control on access to the surrounding forest.

The apple tree development created additional sources for fodder, through the method of intercropping the young apple trees with lucerne. The annual pruning of apple trees provides another source of wood and fodder. The diversity of activities enables households to have sufficient income to purchase consumer goods, such as butane which is used for cooking and domestic heating. These practices result in lower use of forest resources.

The establishment of the Ighrem Association responds to the local need to develop multiple activities. The concept of marketing of the handicraft is the most innovative aspect of this initiative, with a rule to convey one-third of the price to an environmental tax, which is dedicated to reforestation by planting fruit trees in the Douar. In addition to its undeniable social impact, the association contributes to afforestation to compensate for forest harvesting in the future.

Box 6.2 The Agdal system: traditional sustainable land management practice

Populations of these lands practised and still practise sustainable natural resource management. This is the case of the 'Agdal system' prevalent in the High Atlas, which consists of leaving grazing rangeland or forest land fallow during a period of the year, usually spring, to allow plant regeneration. This form of community-based management has demonstrated its effectiveness in terms of the environment, social equity and economy, but is increasingly abandoned in favour of individual modes of land management. This trend has worsened the degradation of the natural resources/environment (Hammoudi *et al.*, 1987 and Auclair and Alifriqui 2013).

In Anzi, the population specialises in handcrafted traditional pottery, iron and wood. Agriculture plays a secondary role in household income generation. The forest plays a key role in the daily lives of the households, for wood crafts and other domestic uses. Institutional innovation has enabled the promotion of products of this wood-working community and improved the incomes and livelihoods of the carpenters. There is, however, concern to ensure that the increased demand does not result in a concomitant pressure on the forest.

Members of the cooperative are beginning to understand the interest in changing the attitude of the population vis-à-vis the forest. Since the identification of this initiative, a partnership project between the cooperative and the Forestry Services has studied how to manage the surrounding forests together and start reforestation and enclosures. Thus, meetings with the DWF of Taroudant took place and an official letter was sent by the cooperative to the Service for the establishment of formulas for concerted management of the forest. In March 2012 there was a public auction of forest products by the DWF. It was negotiated that the sales would be exclusively held for cooperatives and intermediary speculators could not participate. These developments have had a positive impact on the relationship between the Forest Service and local residents of the forest.

Several new practices have been put in place as follows:

• the planting of trees along the river, although the practice remains marginal;
• the purchase of timber which, if developed, could reduce dependence on the surrounding forest;

- some households growing alfalfa on parts of their land to feed their livestock, which decreases the pruning of trees in the forest. Members of the cooperative started the practice of growing alfalfa because it could help to revive their livestock activities, which will have a positive impact on family incomes and simultaneously reduce pressure on the forest;

The cooperative began as a way to circumvent the forest legislation. Since their involvement in the project SCI-SLM, artisans of the Anzi cooperative want to follow the example of their peers in Agouti by integrating environmental concerns into their activities. They envisage a partnership with the DWF to manage the nearby forests, to start reforestation and constructing enclosures. The Moroccan project coordination supports this development in order to change the so-called cat and mouse' relationship, which historically marked the relationship between the local people and forest officers.

The cooperative started with 27 members and currently numbers 37 members. Seven women recently joined the cooperative. They are responsible for painting and polishing the manufactured products. Since the creation and construction of the craft workshop, the horizon of the cooperative has become wider. The brand new building, the pride of the association, will be used for training young local artisans in carpentry and other types of handicraft such as pottery, metalwork and jewelry, as well as for exhibition purposes. This way, the workshop helps young people to create jobs that are sustainable.

People in both communities gradually developed more awareness about the forest and its sustainable management. As a result of their experiences, these communities have more respect for the forest and their environmental awareness has grown. Today they see wood carving as a source of income, which can be accomplished in the context of a 'green economy' based on the sustainability of natural resources. These artisans are building a sensible relationship with the forest and the environment in general. Through their innovative initiatives, these communities are creating environmental awareness with a reinterpretation of ancient practices, such as the revival of the Agdal forest management system.

Lessons learned from Anzi and Agouti initiatives

There are general lessons for both initiatives and specific ones for each of them.

The SCI-SLM initiative operates in a social, economic and environmental context, characterised by comprehensive and evolutionary dynamics. In this context, the initiatives created their own momentum, interacting positively or negatively with multiple factors of the global context and thereby fostering or hampering development.

Examination of both innovations in the context of community conservation management of the forest shows that their effectiveness and sustainability depend on the following conditions:

- the availability of alternative forest resources which decrease dependence on the forest. Alternatives are integrating fodder crops into production systems; pruning trees to use for handicraft; and planting, both on communal lands and private fields;

- the availability of sufficient income to provide alternatives such as butane;
- communities' need to embrace a new environmental culture and search for forms of joint forest management through a partnership with the Department of Water and Forestry.

In the latter case, the main threat would come from neighbouring communities that do not respect the new rules on sustainable forest management. The legal status of forests and user rights of the local communities are a constraint to the implementation of a collaborative management model that will be respected.

Land rehabilitation on the slopes of Ouneine valley

Four brothers and their families, Id Machal lineage, have joined hands in Aït Messoud Douar to reclaim degraded land. They jointly own property consisting of land, water rights and fruit trees, which they have gradually turned into productive irrigated terraces. One of the brothers, who migrated to Casablanca, funded the project. Originally the four brothers had to share the land resources and scarce water with numerous siblings. The family owned a dozen small plots that produced 150 kg of barley annually and an equivalent amount of maize. This was insufficient to meet the needs of its members, which motivated their migration. Their case, which is similar to most of the families in these mountains, expresses the imbalance between a growing population and static natural resources. This initiative on rehabilitating irrigated crop land is aimed at restoring the imbalance between the population and the surrounding natural resources (Figure 6.6).

In 1984 the brothers bought four hectares of uncultivated, degraded and non-irrigated land, with a value of 4000 Dhs[3] per hectare. After rehabilitation, the current value per hectare is between 20,000 and 60,000 Dhs. Currently the family's land covers 20 ha, of which a third is irrigated. The construction work began in the 1990s by clearing the land of stones using a tractor, constructing terraces and digging a well 17 m deep. Initially the land was cultivated for several years with cereals. There is a visible gradation to be seen as one walks upslope – from stony slopes populated by sparse *Ziziphus* bushes, to cross-slope stone lines between which rainfed cereals are grown, to emerging terraces with almond trees along the stone bunds, to rudimentary terraces with some irrigation and finally to fully developed and irrigated benches.

Along this upslope transect it is clear that crop diversification coincides with the development of the land. On the now-mature terraces at the top of the farm, there are various fruit and nut trees, barley, maize, *faba* beans, rosemary, prickly pear (*Opuntiasp*), as well as vegetables (e.g. courgettes, onions, potatoes, tomatoes, eggplant), spices (such as coriander) and medicinal plants. Vegetables were produced for domestic consumption after they tried to sell them in the souk (local market) without much success. New species have been introduced such as mint, wormwood and verbena. Fodder crops are grown including alfalfa, which, when mixed with crop residues of wheat and barley, can feed the animals. Traditional trees such as almond, olive and carob trees are planted to mark the parcel boundaries and used as

Figure 6.6 Trees planted by the association (top) and types of crafts made by the carpenters (bottom)

wind screens. The family community has planted 1,400 trees of olive, carob, almond, peach and prunes. Subsequently, new trees are introduced such as plum, grape, peach, apple, quince and pear. Some of these trees come from Targa-Aide and others have been purchased. Tree products are being sold at the local market.

Irrigation is by groundwater from a shallow well, and a diesel pump is used to lift the water. The family is allowed to use groundwater freely, because it is located under their legally owned land. The initiative in this case is partially social (the idea of a small group of families joining hands to develop land and share certain inputs) and partially technical (diversification into a wide variety of cash crops). It is clearly a community initiative, although a small social entity, and land is without doubt being developed and managed sustainably.

The initiative has multiple economic impacts. First, the land increased in value due to the rehabilitation. Second, the initiative resulted in higher yields of traditional crops such as cereals, the introduction of new crops (fruit trees, horticulture) and the cultivation of fodder to feed their two cows, which provides an extra income of 6,000 dhs per year. The family were able to improve their livelihood and are able to sell many of their new products.

The family rehabilitated degraded, abandoned land by terracing its slopes to prevent erosion. They have afforested the land by planting thousands of trees that produce oxygen and are carbon stocks, with a positive impact on the environment. An important incentive to invest in the land is the fact that the family gained official ownership of the land.

The challenges are a concern about the replicability of the initiative in terms of capital requirements to buy land and install irrigation infrastructures. Second, there is the question of groundwater tables and to what extent groundwater sources are adequate to support expanded agricultural activities. Furthermore, during an exchange visit with farmers from Agouti, it became clear that fruit trees suffered from diseases, which could be treated with pesticides but have negative environ-mental impacts, for example on the bee population. There is also the problem of limited availability of labour. Land and water rehabilitation is very labour-intensive and that explains why these resources may remain underutilised.

Up-scaling towards horizontal spread

This type of initiative has been adopted by a number of families. In Ouneine about 18 plots of degraded land have been transformed into irrigated production systems, as seen in Figure 6.6. Currently these schemes cover an area of 64.5 ha, which is increasing each year.

The Afourigh community participates in the SCI-SLM project with its initiative of a changed management of irrigation water. Its members are adopting the Machal land rehabilitation initiative. In 2012 the family of Aït El Haj in Afourigh started to create irrigated farm land in their Douar. The agriculture here is dry land farming called *bour*, with an area of 6 to 7 ha, much of which had been abandoned by its owners.

Twelve families have plots in this area. These families worked on the land with a tractor together and installed an irrigation system. The Aït El Haj family, who owns most of the land, is financing it. The water is channelled through a 5-km pipe from a storage tank located at Ichmarihn, where families Aït El Haj and Aït Hussein hold units of water.

The creation of these new irrigated lands follows the same pattern of development as the Machal initiative: clearing the plots of stones, installation of traditional irrigation systems or the introduction of new technologies such as spraying or drip irrigation. Water resources come either from natural sources that are adapted or from wells that are equipped with pumps.

Lessons learned from the Machal (Lamhalt) initiative

This innovation is interesting because of its cultivation of formerly degraded land, the development of land resources and water availability, sustainable land management practices and finally the improvement of family incomes. Another advantage is concerned with property rights. Land owners have property rights to the land, and its value has increased tremendously. Their situation was insecure before they started the initiative.

The extension of this innovation depends on the financial means to access and develop the land. The sustainability of these lands will depend on their profitability. Sufficient knowledge of sustainable land management technologies and production techniques ensure high productivity of cash crops.

This is a community initiative involving a small social entity but with a potential for spread, although there might be situations where there is water competition, e.g. when the groundwater table is vulnerable.

New rules for allocating water rights

Irrigated agriculture is a key feature of the Ouneine valley economy. Surface irrigation, from hillside springs distributed by channels into furrows or basins, is the dominant form. However, these springs are fully utilised and water for irrigation has become a constraint, especially during drought periods. The community members described in the Machal initiative have addressed the problem by tapping into the groundwater. However, people whose only source of water are springs emanating from the hillsides have to look for ways to make limited water go further. The case of the five family lineages that comprise the Afourigh community is a fascinating initiative in this regard.

The Afourigh community has an irrigation system that falls into the category of a small-to-medium-scale hydraulic system (Pascon *et al.* 1984, *Question Hydraulique 2*). The system is based on both a physical and a socio-legal principle (Mahdi 1987). Physically the water system comprises a storage basin and a pipe network for water distribution. A transport channel, with water from the source, first flows into the storage pond and then continues with sufficient speed through the network to irrigate the fields. With respect to the socio-legal conditions, the water distribution is divided between 'those with rights', according to well identified water parts; a measured unit of water is called *tiramt*. Units of water are transported through the network of *seguias* (channels), made of soil or concrete, to the plots of the participating owners. The water runs from one plot to another, terrace after terrace, and gradually the channels receive and distribute more water.

This traditional hydraulic system has two disadvantages, in physical and socio-legal terms. Physically there is infiltration of water into the drains, and water is leaking outside the plots, causing huge loss of water. To avoid these losses and to save water, water tower owners decided to fortify the traditional basin, called *Tafraout*, with concrete and they constructed the channel system (*seguia*) in stone and clay by late 1994. The new basin allows larger water quantities and the concrete *seguia* reduces water losses.

The second problem relates to the management of the water. Although generation-old water use rights have officially determined the daily scheduling of the water, the young active farmers within the group have decided to adjust the existing cycle to make the system more efficient. The traditional rotation within the community allowed each family to have half a day (12 hours) of irrigation water, several times within an 11.5-day overall cycle. These water periods are, however, scattered throughout this cycle of 11 days. The result is that families have seen their shares of water indiscriminately scattered over the irrigation cycle. The innovation consists of grouping the water periods to allow owners to use them more continuously (see Tables 6.5a and b).

Table 6.5a Structure of water rights before change

No.	Name of water shares	Time (hours)	Corresponding lineages
1	Ibammochn	24	Aït El haj + Ben Talamint
2	Idihya	24	Aït El Haj + Aït El Haj
3	Aït Makraz	24	Aït Hamou + Aït Hamou
4	Aglagaln	24	Aït Houssein + Aït El Haj
5	Ikakan Zwournine	24	Aït El Haj + Aït El Haj
6	Aït Abderrahman	24	Aït Mohamed + Aït El Haj
7	Boulhssen	24	Aït El Haj + Aït El Haj
8	Tagzzoumt	24	Aït Houssein + Aït Houssein
9	Aït Himi	24	Aït Mohamed + Aït Houssein
10	Akrardene	24	Aït Mohamed + Aït Mohamed
11	Ikakan granine	24	Aït El haj + Ben Talamint
12	Tayassamt	24	Aït Houssein + Jma'a of Douar

Table 6.5b Structure of water rights after the initiative

Family name	Parts water (in tiramt)	Parts water (day)	Parts water (night)
Aït El haj	11	05	06
Aït Houssein	04	02	02
Aït Mohamed	04	02	02
Aït Hamou	02	01	01
Ben Talamint	02	01	01
Jma'a of Douar	01	01 day or night	

Source: Field Survey in 2010.

The initiative was taken by young men from the Douar, after having lived in Moroccan cities and returned to stay in their Douar. Most of them have average schooling grades. There were two periods of negotiation during the implementation of the initiative to change the water rights. First these young innovators needed to convince the owners of the water, namely their parents and uncles and the immigrants who live in the cities of Morocco. The second period of negotiation was when these young people had to reach an agreement on the new rules for allocating the water rights.

In conclusion, the new system has helped to save time and improve water use efficiency, as farmers can plan their workload better. The farmers used to irrigate fields located near the basin, but now their crops can receive ample water at any irrigation application. This can make all the difference in dry years. Moreover, the system is able to store large volumes of water in the basin, which guarantees a consistent water flow with enough speed. The system also reduces water losses caused by evaporation. The increased availability of water after changing the water allocation system has allowed the following take place:

- irrigation of remote plots;
- increasing irrigated areas during drought periods; and
- farmers have been able not only to improve their production, but also introduce a larger diversity of agricultural crops. According to the head of household lineage Aït Hussein, before the introduction of this system, only an area of $200\,m^2$ (equivalent to a fedan) could be irrigated during six hours of closed basin. Currently, it is possible to irrigate four times this area and grow a wide variety of vegetables (tomatoes, onions, carrots, turnips, potatoes, squash) for consumption. In the old system the cultivation of these crops during times of drought was unimaginable. With the new system, even in times of drought, these crops can be grown near the basin.

Up-scaling towards horizontal spread

This community has begun to change the age-old rules for sharing water to increase the efficiency of water management. The importance of the innovation is its impact on a very sensitive issue: the social management of water. This is one of the most difficult innovations to spread to other rural areas, because water rights are considered sacred and untouchable. The hydraulic physical aspects are well accepted and even desired, but touching the private ownership of water is a taboo. From this point of view, the Afourigh community has proven to be avant-garde. The aim of the innovation is to improve the efficiency of water use appropriated by each family, and time savings and improved productivity to follow. It is clear that the implemented water management system is only operational in the dry season or in case of water shortage. When water is abundant in good years or during the rains, there is no need to implement the sharing scheme (Mahdi 1987).

Lessons learned from the Afourigh initiative

Community innovations may affect parts of people's lives that are unlikely to change, such as water rights. The role of the youth has been crucial to evoke this change. The young people have been more pragmatic and they were able to defy the sacred and untouchable character of these rights. Members of all Moroccan communities involved in the SCI-SLM project showed an interest in this innovation. Similarly, members of the Targa-Aide team are fascinated by this innovation. However, it is clear that no community has followed this example nor will any do so in the near future. Some impediments stand against the spread of this innovation and these are discussed below. The challenge of this innovation is how to convince the owners of the water towers to change an ancestral system of water distribution. In the case of Afourigh the elements of success are:

- the small number of owners and family ties and solidarity that exists between them, which facilitated agreements;
- the energy of youth who are living in the Douar, while their elders (parents and uncles) live elsewhere.

Members of other SCI-SLM communities involved deeply appreciated this innovation when they visited Afourigh, but they were certain that its implementation in their Douars is not feasible as the sizes of their communities are much larger and more complex. This makes the consensus around the idea of changing the allocation of the water system improbable. Agouti is an open space where about 80 different families settled and bought land. The social structure of the Douar has become more diverse and more complex. In addition, wealthy families have individual alternatives such as digging wells to improve the quantity of available water.

These assessments lead to the conclusion that it is not enough for an innovation to be considered impressive to be adopted. An innovation is a particular solution to a particular problem. An innovation is an invention of the engineering community that brings a consensual solution, adapted and accepted by the technical and environmental social system.

Conclusions

All four community initiatives identified and characterised demonstrate the innovative capacity and creativity of local communities. These have often been neglected, while they are solutions to address the problem of natural resource degradation. The activities conducted by the SCI-SLM project with these mountain communities living in vulnerable areas helped to stimulate and enhance these community initiatives.

Communities have come to realise the importance of their initiatives and that the role of sustainable land management is crucial to the health of their land. Action research adds value to the innovations and subsequently innovative communities gain confidence in their actions.

Organised exchange visits among SCI-SLM communities within Morocco as well as international exchange visits opened the horizons of the participants to other community initiatives and stimulated them to develop new initiatives, including partnerships with governmental services such as the Forest Department. The SCI-SLM project has created another type of communication between farmers, researchers and, to some extent, the administration around these grassroots innovations.

Since September 2011 a community without an initiative (Douar Tigouliane – CR Tafraouten) participated in various exchange visits organised by the SCI-SLM team. This involvement has had a stimulating effect on its association, Al Mostaqbal, which is engaged in a project to plant olive trees as part of an agricultural development project funded by the Green Morocco Plan.

This experience has shown that some of these Moroccan initiatives have a wide potential diffusion. At the same time, it became evident that, while the interest and importance of a community initiative is clear, its adoption is not always obvious.

The SCI-SLM project has opened up a new perspective to the existing research of Targa-Aide, by putting the focus on initiatives taken by communities themselves. This allowed Targa-Aide to have another look at the communities with whom they have been working for a long time. The community innovations have boosted the vision of Targa-Aide on technical and social dynamics within these communities. Innovative practices were definitely existing, but not really noticed by Targa-Aide, and the SCI-SLM project permitted Targa-Aide staff to discover these initiatives and get involved too.

These community initiatives, however, raise the problem of their sustainability after the end of the project. As Targa-Aide has been working on similar issues for several years, the SCI-SLM approach will be integrated into its agenda and Targa-Aide will try to continue to support these communities.

Notes

1   This chapter has benefited from the collaboration of Ilias Louah and Siham Louriki, former SCI-SLM team members.
2   These are the latest Census of Agriculture data available. Things might have changed, but the figures are for comparative purposes.
3   US $1 = 8.25 dirhams.

References

Auclair, L. and Alifriqui, M. (2012), *Agdal: Patrimoine socio-économique de l'Atlas marocain*. Rabat, Morocco and Marseille, IRCAM, IRD, p. 647.
Hammoudi A., Gilles G. and Mahdi, M. (1987), A Mountain High Atlas Agdal. In *Common Property Resource Management*. National Academy Press, Washington, DC.
Hanich, L., Simoneaux, V., Boulet, G. and Chehbouni, A.G. (2008), Hydrologie des bassins versants du Haut Atlas marocain. Available only at www.iwra.org/congress/2008/resource/authors/abs719_article.pdf (last accessed 19/5/16).
Mahdi M. (1987), Water in Ergulta: Collective management and private rights. In *Common Property Resource Management*. National Academy Press, Washington DC.
Pascon, P. *et al.* (1984), *La question hydraulique*. IAV Hassan II, Rabat, 2 vols.
RGA, (1996), General Census of Agriculture 1996. Ministry of Agriculture and Maritime Fishing, Morocco.

7 Stimulating community initiatives in sustainable land management in South Africa

Avrashka Sahadeva, Maxwell Mudhara, Michael Malinga and Zanele Shezi

Introduction

Unsustainable land management practices are threats to the environment as well as to livelihoods in rural areas of South Africa, where some 17m. people, constituting 1.5m. households, depend on agricultural production, in one way or another. Some 54 per cent of the population of 10m. in the province of KwaZulu-Natal are rural based. Land degradation is reducing the gainful use of arable land, rangeland and woodland during a time of rising demand for food, fibre, fuel, fresh water, fodder, household energy and income (Wessels *et al.*, 2004).

KwaZulu-Natal is one of the four most degraded provinces in South Africa (Department of Environment Affairs and Tourism 2006). Sustainable land management can contribute towards decreasing land degradation. As in the other three partner countries, the Stimulating Community Initiatives in Sustainable Land Management project in South Africa sought to identify ways in which communities are managing their natural resources creatively without significant assistance from external sources.

The four initiatives identified are located within districts of uThukela and uMzinyathi in KwaZulu-Natal. These initiatives are based on management of wattle forest, indigenous forest and livestock grazing management (Figure 7.1 and Table 7.1).

Certain legislation and policies in South Africa are supportive of community SLM initiatives, but others are punitive; this needs to be taken into consideration when working with communities on SLM. It was considered essential that the respective communities understood the legislation governing their initiatives. In this regard, they were linked to relevant government departments which managed the various legislation impinging on SLM by communities. In particular, two National Departments, Environmental Affairs and Agriculture, Forestry and Fisheries, were contacted to carry out awareness sessions on their respective legislation in all four communities. The departments were represented in the National Steering Committee (NSC) and thus were conversant with the SCI-SLM project objectives and approach. The legislation linked to the SCI-SLM initiatives in South Africa and covered during the awareness sessions was as follows:

- National Forest Act 84 (1998)
- National Veld and Forest Fire Act (2001)

- National Protection Areas Act (2003)
- National Environmental Management Biodiversity Act (2004)
- National Action Plan (NAP) (2004)
- Wattle Jungle Conservation Act (2004)
- National Environmental Management Act (2009).

This chapter looks at the four community initiatives and includes the methodology used in selecting and upscaling the initiatives. The first case study describes two closely related community initiatives – the productive management of an invasive tree species in KwaSobabili and the resultant adoption of the same practice by the

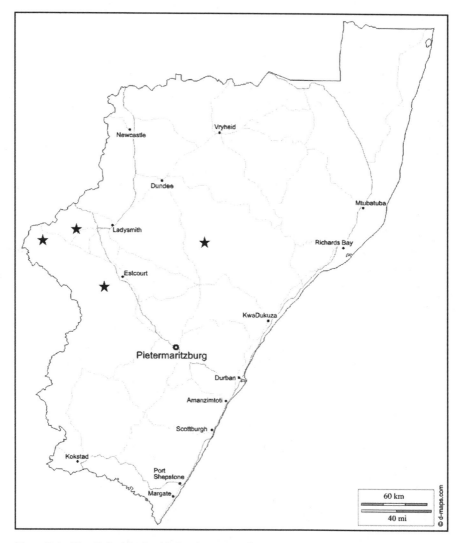

Figure 7.1   KwaZulu-Natal with local municipalities. Stars show the areas in which SCI-SLM community initiatives are located

neighbouring Mathamo Community. The second case study is the rehabilitation and protection of indigenous forest in Gudwini community. The third case study is regarding improved grazing land management and its effects on the control of cattle theft by Amavimbela.

Methodology for selecting initiatives

On the basis of criteria for site selection, the Centre for Environment, Agriculture and Development of the University of KwaZulu-Natal held various meetings with stakeholders, including government departments, academics and civic society organisations, who suggested possible sites in different provinces. CEAD also used various networks and platforms (such as Sivusimpilo-Okhahlamba Farmers' Forum), as well as PROLINNOVA[1]-related projects that dealt with the linkage of innovations to knowledge and understanding.

In addition to SCI-SLM's selection process of identifying a community initiative, other criteria were included. These criteria were adopted on the basis of experience of other projects involved in identifying community innovations (such as PROLINNOVA and FAIR[2]), as well as those set out in the SCI-SLM training. These criteria highlighted three particular prerequisites: (i) that the initiative was locally developed, (ii) that it benefited the community and (iii) whether it was new or an improvement on existing practices. The initiatives identified were characterised using the SCI-SLM methodology and a summary is presented in Table 7.1.

Case studies

Productive management of invasive tree species: KwaSobabili community initiative and spread to Mathamo

During the period from the late 1990s to the early 2000s, South Africa experienced a high rate of HIV/AIDS-related deaths, which led to more funerals than normal. It is part of Zulu culture to use logs during burial to separate the soil from the body: these are termed 'burial logs'. This increase in deaths resulted in increased use of burial land and wood for burial logs. This led the KwaSobabili community to take action to protect their forest. The community realised the importance of the wattle forest as a source of burial logs. Since 2000, 65 ha of wattle trees have been protected and managed by the KwaSobabili community of 350 households.

Box 7.1 *Acacia mearnsii* (Black wattle)

Wattle trees were imported from Australia for industrial use and were planted in restricted areas, but spread into the wild and thus became 'thirsty' invasives in water-scarce South Africa. However, communities used the trees for day-to-day household needs such as firewood, cattle *kraal* construction and house building. During the mid-1990s, the Working for Water (WfW) programme started a programme to remove invasive species, especially wattle.

Table 7.1 Overview of the four community initiatives in KwaZulu–Natal, South Africa

Initiative	1. Wattle forest management	2. Wattle forest management	3. Grazing land management	4. Indigenous forest management
Name and location of community	KwaSobabili, Imbabazane Local Municipality, uThukela District	Reserve, Okhahlamba Local Municipality, uThukela District	Okhahlamba Local Municipality, uThukela District	Gudwini, Msinga Local Municipality, uMzinyathi District
Year started	2000	1992	2000	1945
Origin/source of community initiative	Modified tradition; community members have taken initiative, which is recognized by the chief	Modified tradition; community members recognised the need for the forest	New social arrangement of a core group comprising UKhahlamba Livestock Cooperative (ULC) members	The initiative was started by the chief. It became strong in Gudwini through a new social arrangement
Motivation/trigger of initiative	Need for burial logs (logs are traditionally laid in a grave to separate a body from the earth) and timber for construction	Need for fencing, burial logs, firewood and construction	Progressive increase of cattle theft in the region and limited support from law enforcement agencies. Thus, fewer cattle brought by owners to the mountains where the rotational grazing camps had been established. Overgrazing of pasture close to homesteads due to high concentration of cattle in a restricted area	Most of the indigenous forests under the Traditional Authority were over-exploited, and degraded. Use not regulated. In addition, need for protection from outsiders as the forest is an asset and source of wood, fruits and indigenous medicinal plants

Table 7.1 Continued.

Initiative	1. Wattle forest management	2. Wattle forest management	3. Grazing land management	4. Indigenous forest management
Community initiative: brief description	In 2000, 65 ha of wattle trees were protected and managed by a community of 350 households	Community is residing on land previously owned by a community commercial farmer who produced wattle. The land, including 45 ha of wattle forest, was given to the community. Thus in 1992, the sub-committee in Reserve B protected the wattle forest for community use. Sub-committees in other use areas (Reserves A and C) did not adopt the same practice	Before formation of AmaVimbela, owners did not allow their livestock to graze far from residential areas because of fear of theft. This had a negative impact on land as there was overgrazing, resulting in crop production being compromised. The extra protection to cattle under AmaVimbela allowed owners to send their cattle far away, thus making full use of the rangeland. This has led to sustainable use of the land, and minimized soil erosion through driving the cattle to and from the grazing lands	The community is preserving and managing its indigenous forest. The village head works with a few elected village policemen to ensure that only dead trees are used for firewood and that trees are not cut for commercial use outside the community. Generally, the forest provides for household uses: wood (source of energy), wild fruits (though minimal), logs for fencing fields and buildings
Area under SLM	65 ha	Area under cultivation is 40 ha	3,500 ha	45 ha
Initiative typology	Social and technical	Social and technical	Social	Social and technical
SLM main technology category	Community forest management (CFM) through productive management of invasive wattle forest	CFM through productive management of invasive wattle forest	Grazing land management through opening up of grazing land that had previously not been accessible because of livestock theft	CFM through forest protection

Management has been carried out by a committee that ensured that wood was not stolen. The dominance of males in the committee in such a culturally rooted community allowed the committee to impose a restriction on access to the forest resources. The chief allowed the poorest community members to use the wood for household uses and burial logs. This is a unique initiative where the community protects a forest, while surrounding communities used their forest unsustainably. KwaSobabili has developed rules and regulations on the use of their forest. The rules have been beneficial as its members are able to collect and use forest resources to meet daily needs for energy, shelter and burial logs. This allows members to save money. The evolution of the initiative is presented in Table 7.2.

Description of the KwaSobabili forest initiative

The SCI-SLM project came across KwaSobabili wattle forest management after holding various meetings with people within the community: the tribal secretary, the environmental committee and KwaSobabili wattle forest patrollers. These meetings pointed out the importance of the wattle forest to the community for the use of burial logs, firewood, construction of houses and livestock pens, and craftwood. Using the project characterisation criteria (see Chapter 3), the initiative was originally identified as a social initiative. However, through the linkage and interaction during the projects and relevant stakeholders, the initiative was classified as simultaneously a social and a technical initiative (Figure 7.2).

Figure 7.2 Community members working in the forest after inception of project

Table 7.2 Timeline of the important changes and evolution that has occurred in KwaSobabili wattle forest community initiative

Year	Description of evolution
1939	The South African *apartheid* government, through the Land Act of 1913, imposed the Betterment Scheme in 1939. The scheme forced removal of black people from their farms onto reserves. Their land was demarcated into grazing land and arable lands, resulting in a change of usage and lifestyle.
1994–2000	Land was shared amongst all subjects under a chief. Forest communities were given a forest to use for household and cultural purposes. KwaSobabili community was the only community that took up the initiative to maintain their forest in early 2000.
2009	The SCI-SLM project identified the community initiative. The project team visited the area and met the chief, tribal council and community members.
2010	Meetings identified that only men and one woman were part of the committee working on the initiatives. Through project interventions, the community selected a new committee which had equal numbers of men and women. More community meetings were held in relation to development and identifying areas of intervention for the forest. Through the project, a total of 250 members work in the forest to clear unwanted and old wattle for community use.
2011	UKZN took the community members for a cross-visit to Natal Timber Extract in Greytown to observe the technical aspects of maintaining a wattle forest and how to convert their wild forest into a manageable forest – away from vital water sources. As a result, the forest committee allowed community members (who worked in the forest) to collect wood for winter fuel; those who did not work had to buy the wood. This was the first time that this had happened since the community started managing the forest.
2012	UKZN linked the community to the Department of Agriculture, Forestry and Fisheries (DAFF) to register the forest and provide technical and financial assistance. In turn, DAFF linked the community to WFW, which funded the community to clear weeds; this resulted in job creation for community members. The Department of Environmental Affairs (DEA) provided an awareness session for the community on community-based natural resource management. Forest registration for permission to plant and keep a wattle forest was granted with assistance from SCI-SLM project team and DAFF. A community member attended and participated in the third project steering committee meeting in Ghana. The community hosted a field day attended by neighbouring communities.
2013	Committee skills training was conducted by the SCI-SLM project team. The committee met again to review their roles on the basis of what they had learnt during the training. The community stopped using a piece of land, which will be used to plant wattle seedlings. The project identified a neighbouring Mathamo community, which then adopted the initiative in KwaSobabili. The neighbouring Mathamo community want to replant and rehabilitate their forest for community use.

Initial interviews indicated that KwaSobabili already had an existing seven-member committee, which included one woman and a representative of the tribal council. The committee met once a month to discuss the forest activities. The committee reported to a community meeting where the forest was discussed. Originally the KwaSobabili forest was only harvested by the poor with permission from the tribal council. As a result of the introduction of the SCI-SLM project in order to improve and strengthen the initiative, the community included more women on the committee. Since 2011 wood has been harvested every winter by everyone in the community (which was the first time in the community) in order to control and manage the spread of the forest. The project undertook a number of activities to improve the initiative. First, a committee skills training (in terms of member structure and roles) was conducted. SCI-SLM also facilitated a Working for Water (WfW) initiative, a government programme for the identification and removal of alien invasive species close to water sources, which was implemented in the community for three months in 2012. The latter employed community youths to work in the forest, thus resulting in income generation, employment and enhancing their involvement.

At the inception of the project, there was an indication that KwaSobabili lacked technical skills to manage the forest, so they resorted to a blanket restriction on harvesting the trees, except for selected households. Through the interventions of various stakeholders through SCI-SLM, the forest was divided into segments to allow 'rotational harvesting', where community members harvested one part at a time to allow for younger trees to establish. WfW provided chemicals and equipment and then employed community members to thin and maintain the forest. In addition, the SCI-SLM project arranged for the group to go for a cross-visit to Natal Timber Extract (NTE) (see Table 7.1), where they were exposed to the effective management of wattle forests (land preparation of forest sites, seed storage, planting of seeds, monitoring the tree growth). They also learnt better coppicing methods as well as encouraging better forest management through rules and regulations, appropriate thinning of trees and controlling of an invasive exotic blackberry (*Rubus fruticosus*), preventing grass from encroaching and ways of reviving unmanaged wattle forests. In addition, the groups also learnt about using branches harvested from the forest for fencing to prevent livestock from damaging the trees.

Added value of SCI-SLM to the initiative

One aim of SCI-SLM was to strengthen community initiatives. As part of the broader SCI-SLM methodology, the project identified the community and worked with it to improve and strengthen it through the following steps:

- The community members were introduced to NTE in Greytown. This resulted in activities spearheaded by the forest patrollers working with community members, as discussed and advised by NTE during the cross-visit. Community members became aware of the need for pruning, cutting and replanting of wattle forest.
- They were linked with stakeholders such as the DAFF and WfW. The involvement of these agencies in the initiative facilitated awareness by the community

of requirements of the Wattle Jungle Act, such as converting an unmanaged wattle forest into a managed one.

- The community developed a plan to maintain the wattle forest. This plan was conducted in demarcated areas (i.e. harvesting, clearing, seed scarification and replanting).
- WfW provided technical support to enable the community management of the forest and the clearing of invasive species. This created employment opportunities within the community as 12 community members were employed on a part-time basis for three months.
- All wood harvested through thinning the forest was sold to community members. The money was then used for maintaining and developing the forest.
- Provision of a platform for women's participation and training on committee and management skills made the committee aware of individual roles and how to minimise abuse of power. Women's participation in the committee increased.

The SCI-SLM project helped the KwaSobabili community understand the value of their wattle forest. The community now associates the forest with creation of employment opportunities and potential for saving energy costs (a total of approximately US\$ 7,500 per annum from the whole forest). And members were allowed to harvest wood in a managed manner. These benefits increased community cohesion. The community members are continuing to review their rules to ensure maximum benefits are derived, while the resource is sustainably managed. Recently a rule was implemented which stipulated that, if members worked in the forest, they could collect wood for free during the winter period.

Because of savings that accrued indirectly to community members, their participation in protection of the forest increased. This resulted in surrounding communities approaching them to find out about the initiative. The interest in the initiative by the neighbouring communities led to the SCI-SLM project facilitating a field day in November 2012, attended by other communities. The attendees included government representatives, members of the traditional council, community members from Reserve B in Okhahlamba Local Municipality of Bergville and a neighbouring community which also adopted a similar initiative in sustainable land management involving wattle forests.

The field day at KwaSobabili led to the following actions (Figure 7.3):

- increased female participation;
- awareness of the initiative by the local municipality;
- awareness raising of the initiative in neighbouring communities; and
- a neighbouring community of Mathamo adopting the initiative.

The field day raised awareness of the initiative among the communities that had potential to adopt the sustainable management of forests, such as Reserve B and Mathomo. Even though the Reserve B community already had a similar wattle forest initiative (Table 7.1), it was experiencing challenges of how to organise members to manage the forest. The field day allowed members from Reserve B to observe how KwaSobabili community worked and thereafter improved the management of its own forest. As a result, the Reserve B community strengthened its initiative. In addition, two communities neighbouring Reserve B, i.e. Reserves A and C, also started proactively managing their forests. The SCI-SLM project is in contact with Rural Forest Management (RFM), a private company that assists rural communities in setting up business plans for the production of wattle, to link them to the KwaSobabili. It is hoped that this linkage will eventually translate into employment creation within the community.

Spread – horizontal upscaling

Mathamo has 70 ha of gumtrees (*Eucalyptus sp.*) and a wattle plantation which has not been well managed. As a result of poor forest management, the forest has declined in size. The community had to buy wood from neighbouring commercial plantations for burial, household use and other uses.

After the field day, the Mathamo community selected a committee to develop the forest. They then approached KwaSobabili to understand the methodology and adapt it to their current circumstances. The Mathamo committee decided that they would use the forest for burial logs, construction poles and firewood. The youth seem to be more active in the forest maintenance and upkeep with the hope that they would secure employment.

Figure 7.3 Field day at KwaSobabili

Indigenous Forest in Gudwini

South Africa is not well endowed with indigenous forests. Large parts of the forest under the Tribal Authority are under threat as they are neither conserved nor regulated. Many rural communities primarily depend upon and use indigenous forest resources to meet daily needs for energy, shelter, food, burial logs and medicine. This in turn allows scarce financial resources to be used to access other household needs and the accumulation of assets necessary for a secure livelihood. Without these forest resources, communities would lose part of their heritage and would be forced to look elsewhere for such resources. On that basis, Gudwini forest is amongst the few intact forests (Figure 7.4). Chief Simakade Mchunu initiated the preservation of forests in the Machunwini area in approximately 1945 through the tribal council (Table 7.3).

Table 7.3 Timeline of the Gudwini community initiative showing the important changes and evolution that occurred before and after the SCI-SLM project

Year	Description of evolution
1945	All areas in Machunwini revived the local community management (conserving) system. Local men and women worked as forest patrollers who enforced the rules and regulations and organised regular community meetings to further discuss forest issues.
1945–2005	Many challenges were faced within and between communities e.g. a high level of theft of trees from the forest area; recovery after being damaged is very poor for certain tree species, which led to the degradation of indigenous trees.
2005	It became evident that Gudwini community area was more active in sustaining the forest. This was headed by iNduna and local men as forest patrollers who enforced the rules and regulations and organised regular community meetings to discuss forest issues.
2010	The SCI-SLM project identified the Gudwini community initiative and started stimulating the community to improve the initiative.
2011	SA SCI-SLM project team facilitated the exposure visit for Gudwini to Amazimeleni community in the Ongoye area of KwaZulu-Natal. This visit dealt with the management of nurseries for indigenous trees. The village head of the community participated in the SCI-SLM second project Steering Committee meeting in Uganda.
2012	Community members collected seeds and planted these at homesteads. The SCI-SLM project linked the community to key stakeholders such as DAFF, Ezimvelo KZN Wildlife, Msinga Local Municipality and Provincial Department of Agriculture and Environmental Affairs. Awareness sessions were conducted in community-based natural resource management sessions.
2013	Community demarcated land for transplanting seedlings that they had planted and raised. Members of the Gudwini community were taken to Silverglen Nature Reserve Nursery in Durban, KwaZulu-Natal, where they were shown various indigenous trees and how to raise these from seeds.

Figure 7.4 Rehabilitation of indigenous forest in the Gudwini community initiative

Generally, the forest provides resources used at household level, e.g. firewood, wild fruits (though minimal), logs for fencing fields and for constructing houses. Economic benefits such as craftwork, use of medicinal plants and, more importantly, grazing of livestock are also advantages.

Current status and cultural significance of the forest

Community members use their forest for different purposes. However, there has been a decline in availability of resources for extraction due to overuse, illegal extraction and outright theft. Many tree species are becoming endangered. Thus it became imperative for the community to protect what was left (Table 7.4).

One of the many uses for trees is for firewood. Culturally, it is expected that each married woman would have a pile of wood called *Ibonda* or *Isithondo* (meaning 'together'). The pile is used culturally to demonstrate that a woman is present in the household. This practice tends to exaggerate the true demand for firewood from the forest. Although all community members have equal access to the forest resources, women are the main beneficiaries.

Description of the forest initiative

This community initiative was identified by SCI-SLM as a social initiative. A local organisational structure consisting of a working group of 15 people, who are

Table 7.4 Indigenous tree species that have cultural value to communities and/or are used for firewood

IsiZulu name	English and common name	Cultural description	Status
Msululu	Euphorbia tirucalli or 'milk bush'	When dried can be used for preparation of Zulu beer. It is used traditionally for cleaning of goats' eyes.	Some members plant the tree at household level.
Isiphapha	Euphorbia triangularis	Bees collect pollen from its flowers.	Few trees left in the forest. Some members plant at household level where seedlings are raised.
Mnquma	unknown	Produces a sweet juice which can be drunk by humans and birds. It is also used for craft work.	Few trees left in the forest. Some members plant at household level.
iDungamuzi	Scolopiazeyheri sp. or 'thorn pear'	Not used for domestic purposes.	
umPhumbulu	Mimusops obovata or 'red milkwood'	Used for fruits and firewood.	Few trees left in the forest. Communities having difficulty germinating seeds.
umVithi	Boscia albitrunca or 'shepherd's tree'	Used as building materials for roofs.	
Mqathathongo	Premna mooiensis	Used for fruits and firewood.	Some members plant the tree at household level.
Ilotha		Used as remedy for skin rash.	Some members plant at household level.
uMkhukhuze	Cassineperagua or 'cape saffron'	Has fruits; the whitish tree used for constipation.	Few trees left in the forest. Farmers raise seedlings at household level.
Umdolofiya	Opuntia sp or 'prickly pear'	Has edible fruits.	Neither endangered nor extinct.

Note: Zulu names were translated into common English names using Pooley (1993).

involved in patrolling and rehabilitation of the forest, manage the forest. Seven committee members have specific functions. The group is dominated by women (14 of 15) and the only male representative is the village head. Forest patrollers work on a voluntary basis. They monitor the harvesting of trees to ensure that trees are not overexploited. Community members collect seeds, especially of species that are now poorly available in the forest, and plant these.

The Gudwini initiative has contributed to better regulation of forest use, including the type of trees that can be harvested sustainably. Through various meetings, it was found that there were many challenges such as management and technical skills on how to sustainably manage the forest; soil erosion; termites; certain species failing to re-germinate; as well as theft of timber. In order to address these challenges, communities hold monthly meetings where forest management is also discussed. They also charge fines to those illegally cutting trees.

Technical skills to manage the forest were lacking and institutional building was needed. In order to address the technical challenge, the need for an exposure visit was identified as one effective way for communities to learn from other rural communities. Hence the community went on a cross-visit to Emazimeleni Community in Ongoye area within KwaZulu-Natal province, organised by SCI-SLM. They also visited Silverglen Nature Reserve nursery in Durban, KwaZulu-Natal. This nursery is involved in protection and planting of a wide variety of indigenous plants. The technical skills that communities required were knowledge about tree planting.

Added value from SCI-SLM

Various sessions were used for making communities aware of the importance of the forest. More women and youth participated in these activities. Since the harvesting of wood and firewood is free of charge, this translated into an equivalent saving of US$ 800 per annum per household.

Evaluations suggested that the SCI-SLM exposure sessions and cross-visits benefited communities in the following ways:

- Collaboration with other communities was positive for sharing information and rules and regulations.
- Exposure visits allowed people to gain knowledge about seed propagation.
- They also learnt about:
  - techniques for planting of seeds and seedlings, when to collect seeds to allow indigenous seeds to germinate;
  - different seed treatments to improve germination for various types of seeds;
  - types of soil for different seeds; and
  - the depth and amount of light exposure for seeds and seedling planting.

Grazing land management in Amazizi and Amangwane traditional authorities

Land degradation is one of the most severe and widespread environmental problems in South Africa, and it affects food security (Wessels *et al.* 2004). Improvement in grazing management increases forage productivity and the quality of land. Standard, continuous stocking fails to produce the maximum amount of forage and results in overgrazing and land degradation. This can be addressed through controlled livestock grazing. However, another reason for overgrazing and land degradation is livestock theft, which has led to grazing being confined to safer land, close to residences. The creation of Amavimbela is an innovative effort by livestock owners to solve the livestock theft issue, while indirectly addressing land degradation and overgrazing issues in the Okhahlamba local municipality.

Okhahlamba is one of five local municipalities in the uThukela District. The municipality contains the Drakensberg Mountains, a World Heritage Site. The mountains are a physical border between Okhahlamba (KwaZulu-Natal), Free State Province and Lesotho. Several major South African rivers have their sources within these mountains. Okhahlamba region is at present home to three traditional authorities, the AmaNgwane, AmaZizi and AmaSwazi, as well as commercial farmers who came into the area during colonial and apartheid periods. Due to its location, the region is exposed to cross-border criminal activities. Cannabis trading is common and this leads to other illegal activities such as livestock theft. Livestock theft has been rife in the rural communities, which has prompted different communities to develop strategies to combat the crime and protect their livestock. A strategy of some individual livestock owners has been to herd their livestock as groups. They even sleep in the grazing areas to ensure that their livestock are not stolen at night.

Amavimbela community initiative

Amavimbela (translated as 'the protectors') Network operates in the Northern Drakensburg Mountain in uThukela District, KwaZulu-Natal province. The network is a sub-structure of the OLC. It is semi-autonomous, with its own executive structure that regulates its operations. The body of membership is drawn from 15 villages in the OLM. Community members monitor cattle movement and recover stolen animals in the region to reduce theft and the insecurity that this brings. In addition, Amavimbela aims to develop and strengthen local institutional structures, resulting in the effective use of communal grazing and rangeland improvement. Through the system, the network encourages cattle owners and herders to identify areas to graze cattle without restricting the livestock to pasture land close to homesteads. Though livestock is the core function, the work of Amavimbela has extended to assisting households recover other stolen goods because of increasing theft in the area.

The group originated in 2000, from a former group called Scorpion. The former group used violence to recover stolen cattle and had been infiltrated by criminals. Some members of Scorpion were cattle thieves, which compromised its integrity. Contrastingly, Amavimbela is registered with the South African Police Services (SAPS) – and is acknowledged by the Traditional Authorities. The committee recruits members to Amavimbela. Each community represented in Amavimbela has a local committee of five members. Female members are included in the committee for strategic purposes because women can act as informers when they come across livestock that is stolen. Each community member who wants livestock to be protected by Amavimbela pays a membership fee of US$ 1.00 annually. Subscriptions are used for funding the operations of Amavimbela during meetings and recovery of stolen livestock.

Amavimbela recovery process

At community level, Amavimbela undertakes an initial investigation once stolen livestock have been reported. Since Amavimbela works as a network, members from one area will inform members in others. Therefore, for the system to work, mobile phone communication is used. During the search, Amavimbela request the person herding livestock to produce a Livestock Removal Certificate, which should be stamped and signed by the village dipping leader of the area where the livestock is claimed to originate from. Without such proof of documentation, suspects are handed to the police. The police charge the culprit accordingly.

Contribution of Amavimbela grazing land management

Cross-border cattle theft between South Africa and Lesotho has existed for many years but it has not been treated as a priority crime. Amavimbela observed that livestock theft always goes hand in hand with cannabis trade between the two countries. As a result, livestock owners preferred to graze their livestock on safe pastures near their homesteads, rather than on the mountain rangelands. These practices have put a strain on the grazing land closer to the homesteads, which resulted in degradation and the risk of livestock devouring farmers' field crops and vegetables. Traditionally, smallstock (goats and sheep) are tethered close to the homesteads.

Amavimbela has reduced cattle theft, and the co-operation between law enforcement agents and the community has improved. However, it is not easy to find backing statistics from the police and thus to credit the reduction in thefts to Amavimbela. The work of Amavimbela has 'freed the rangeland' and helped to resuscitate land close to the residential areas for agricultural production.

SCI-SLM and Amavimbela

After conducting focus group discussions in 2011 followed by another in 2013, to understand the activities of Amavimbela, the following changes in the association were attributed to SCI-SLM:

- Amavimbela increased their female participation from 26 per cent to 30 per cent.
- As the result of working with Amavimbela and providing exposure visits in relation to grazing land management, three of the 15 villages have established rotational grazing systems, which enhances the benefits of livestock management.
- There is better understanding of natural resource management and grazing systems.
- As part of their role as a livestock association, Amavimbela have introduced livestock auctions for smallholder farmers. This is a good income generator for livestock owners.
- Farmer-led documentation awareness was conducted with groups during meetings. This has resulted in the groups buying cameras to document the livestock recovery process and other land management practices.

Conclusion

There are many interventions that can be attributed to SCI-SLM. In the South African cases, identified communities have been practising their own initiatives such as grazing land management and afforestation to combat land degradation and contribute to improvement of livelihoods. Communities practice their own initiatives that contribute to effective SLM whilst unknowingly following certain rules and regulations set by government on improved SLM approaches. Because of the engagement with the project, communities have been sensitised in knowing the policies that govern and protect their approaches, provided that they are registered.

In terms of overall experiences, various key findings were identified. For example with respect to the two wattle forest initiatives, the communities have groups who work in the forest and ensure that wood is not stolen. There are policies in place to control and regulate the uncontrolled spread of the 'thirsty invasive' wattle. In some cases government can remove wattle if no value is attached to it; however, with the Wattle Jungle Act of 2004, communities can register their forests and manage these to prevent uncontrolled spread to water sources. Since the initiatives identified manage and control the spread of wattle, the communities registered the forests to ensure that they could manage them legally. It was evident that unmanaged forests became depleted and their grazing land and forest area was invaded by alien plantation and weed. The lessons learnt from wattle initiatives were as follows:

- Community working groups need to be established to ensure that rules and regulations are followed.
- When communities work to manage and maintain their forest and assign value to the forest, they will have sustainable supplies for years to come.
- If the forest is well maintained, the community members save money, which may ultimately be invested in agriculture.
- Communities can register a forest to ensure that they have advisory and technical assistance from government.
- The initiatives can be used as an example for other communities to adopt similar initiatives.
- More technical skills on invasive weeds need to be taught to members of communities.

In terms of indigenous forests, Gudwini forest is amongst the few in KwaZulu-Natal that is still basically intact as the community preserves and manages their forest. Being able to collect and use forest resources to meet daily needs for energy, shelter, food, burial logs and medicine allows scarce cash resources to be used to secure other household needs and the accumulation of the necessary assets for a more secure livelihood. This includes education of children, investment in agricultural tools and capital for income generation activities. Many rural communities primarily depend on indigenous forest resources and, without these resources, communities will lose part of their heritage and will be obliged to think of alternatives. In addition, they are contributing to SLM and to the protection of endangered indigenous species. Working with the Gudwini community, the following was realised:

- Communities have values associated with the forest.
- Protection of the forest requires that communities are made aware of the policies and laws that apply to their resource.
- The government needs to hold more awareness days with the involvement of communities.
- Awareness raising needs to be carried out at all levels of society, including amongst the youth and elderly.
- Education of the population starts with the youth, and educational materials should be distributed to schools and tertiary institutions.
- Assistance by government departments as lead agents in conveying the importance of indigenous trees and involving society in campaigns, such as Arbor Week, is very important.

In terms of grazing land management initiatives implemented indirectly through Amavimbela, this initiative was unique in that the group was unaware of the positive effect they were making towards grazing land management. It suggests that, if government work closely with communities, the results can be mutually beneficial. The main outcomes can be summarised as follows:

- Community initiatives exist in land management of grazing lands.
- Community initiatives reduce overgrazing of land, which in turn prevents degradation.

Communities are a critical component for achieving SLM. There is a need to stimulate the initiatives so that they are improved and out-scaled. The development of policies that impact on resource use should begin with consultations at community level to ensure both relevance and adoption.

Notes

1   PROLINNOVA (PROmoting Local INNOVAtion in ecologically-oriented agriculture and natural resource management) is a global partnership that promotes local innovation amongst individuals and communities (www.prolinnova.net).
2   Farmer Access to Innovation Resources (FAIR) was a project that supported community innovativeness by making available funds for improving innovations. It was operated through PROLINNOVA.

References

Department of Environment Affairs and Tourism, (2006), *South Africa Environment Outlook. A Report on the state of environment.* Department of Environmental Affairs and Tourism, Pretoria.

Pooley E. (1993), *The Complete Field Guide to Trees of Natal, Zululand and Transkei*. Natal Flora Publications Trust, Durban.

Wessels, K.J., Prince, S.D., Frost, P.E. and van Zyl, D. (2004), Assessing the effects of human-induced land degradation in the former homelands of northern South Africa with a 1 km AVHRR NDVI time-series. *Remote Sensing of Environment* 91, pp. 47–67.

8 Community initiatives for improving degraded ecosystems in Uganda

Stephen Muwaya, Richard Molo, John Ssendawula, Swidiq Mugerwa, Alex Lwakuba and Sabina Di Prima

Introduction

In Uganda, land degradation is a widespread problem exacerbated by population pressure[1] and the use of unsustainable farming practices in already vulnerable soils. It has been estimated that in some regions 60–90 per cent of the land area is affected by soil erosion (NEMA 2005). Furthermore, nutrient mining makes the situation worse as nutrients are lost and not replenished. The use of chemical fertilizer is generally very low.

Land degradation directly affects the livelihoods and resilience of the predominantly rural-based smallholder producers.[2] However, it has more far-reaching impacts. Land degradation is acknowledged to be the major impediment to sustainable growth in agriculture, natural resource productivity and the economic development of Uganda (MAAIF 2010a). Drechsel *et al.* (2001) estimated that the cost of land degradation amount to 6–11 per cent of Uganda's agricultural gross domestic product (AGDP) annually. Other studies have estimated that soil erosion alone accounts for over 80 per cent of the annual cost of environmental degradation (NEMA 2005). Also, the cost of soil nutrient loss, due primarily to soil erosion, was estimated to be about US$ 625m. per annum based on 2001/2002 prices (NEMA 2005).

To address the complex and multi-faceted challenge of land degradation, in 2010 the Government of Uganda developed the Strategic Investment Framework for Sustainable Land Management (SIF/SLM) and integrated it into the Agricultural Sector Development Strategy and Investment Plan (DSIP) and the National Development Plan (NDP) (MAAIF 2010b; NPA 2010). The Strategic Investment Framework was also developed to contribute to the regional initiatives to scale-up sustainable land management in Sub-Saharan Africa (e.g. the World Bank-driven TerrAfrica and the Comprehensive African Agriculture Development Programme – CAADP).

In Uganda, the Strategic Investment Framework targets specifically the so-called 'land degradation hotspots' identified on the basis of biophysical and socio-economic factors (Voortman *et al.* 2000; World Bank 2008). These hotspots include the Cattle Corridor, the eastern and south-western highlands, the Lake Victoria crescent regions and the agricultural landscapes of eastern and northern Uganda (see Figure 8.1).

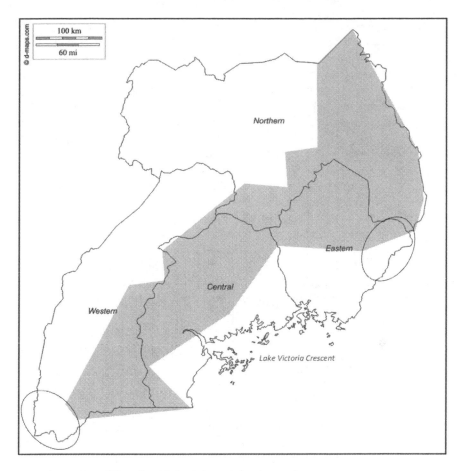

Figure 8.1 Map of Uganda with land degradation hotspots

The Cattle Corridor is a dryland belt stretching from the north-east to the south-west of the country. It is one of the most fragile ecosystems in Uganda, characterised by shallow soils, poor agroecological potential, low population density and poor market access. The region registers low and erratic rainfall, recurrent droughts and high temperatures. Extensive agriculture is commonly practised but seldom combined with adequate conservation measures. Land use trends show an increase in deforestation, overgrazing and bush burning, as well as a reduction of fallow periods. As the name implies, this region is more suited to livestock than crop production. However, the rise in food demand is leading to the expansion of agriculture. For instance, in the central districts of the Cattle Corridor, land was predominantly used for communal grazing and as a transit corridor for livestock. However, recent trends show an increase in individual tenure rights and the use of fencing and paddocking to limit the access and movement of livestock.

The south-western part of the Cattle Corridor is characterised by undulating steep hills which experience serious land degradation due to overgrazing, frequent bush burning and inappropriate farming practices. These activities pose a serious risk of soil erosion, with accompanying wide-ranging effects including flash floods, landslides, gullies and the destruction of houses and roads. The hilltops are communally used for livestock grazing, while the lowlands and valleys are used for settlements and agricultural activities (typically the cultivation of coffee, bananas and annual crops). The high population growth, increasing migration and settlement in these areas is leading to gradual land fragmentation, reduced average land holding per household and reduced fallow periods. Most of the land is managed under a customary land tenure system, where acquisition is through inheritance or purchase via local agreements. Large portions of land are public, owned by government, and are under various forms of use, for example forest reserves.

The diverse types and changing dynamics of land tenure systems in the Cattle Corridor have contributed to insecurity of land access, especially for pastoralists and smallholder squatters. Furthermore, the counter-productive land tenure systems, ranging from customary public land to leasehold and '*mailo land*',[3] have been blamed for poor land management and increasing land degradation (NEMA 1995). These conditions make the population in the Cattle Corridor extremely vulnerable to the impacts of land degradation – thus it is a priority hotspot in Uganda.

History of community initiatives in Uganda

In the arduous fight against land degradation, Uganda has realised that one of the most effective solutions comes directly from the local communities and their innovations. After 25 years of field experience, there is now recognised evidence that community initiatives (technical and/or social) are a key entry point to sustainable land management, delivering long-term impact, even in the most degraded landscapes.

The piloting of community initiatives started in early 1994 with the Conserve Water to Save the Soil and the Environment (CWSSE) project, managed by Silsoe Research Institute and funded by the UK Department for International Development (DFID). The CWSSE project was implemented in the hilly terrain of Kabale District, in south-western Uganda. As well as making an inventory of indigenous soil and water conservation (SWC) practices, the project studied two of these (banana mulching and trash lines) in detail. Furthermore networks of 'farmer-researchers/ innovators' were set up (Critchley *et al.* 1999a). In 1998 a second phase of this initiative was established under the Indigenous Soil and Water Conservation project (ISWC) funded by the Netherlands Government, whose objective was to document the extent of traditional systems in Africa. In addition to promoting best practices, ISWC supported farmers to participate in study visits which, in turn, encouraged the testing and adoption of further land husbandry ideas (Reij and Waters-Bayer 2001).

Building on the CWSSE/ISWC experience, another project focused on community initiatives was launched: Promoting Farmer Innovation. The PFI project (1997–2001) operated under the framework of the United Nations Convention to

Combat Desertification (UNCCD). It had a regional approach, covering Kenya, Tanzania and Uganda, and it targeted specifically the drier areas, prone to drought and desertification (Critchley *et al.* 1999b; Mutunga and Critchley 2001). PFI's entry point was the identification and characterisation of farmer innovators. The project developed and promoted the institutionalisation of a methodology for assessing the impact of farmer innovations and the effect of the exchange visits involving farmer innovators. The outcome of these pilot initiatives (CWSSE/ISWC and PFI) together with a later, related initiative, termed PROLINNOVA[4], was evaluated in December 2000 during a joint regional review meeting with the idea of developing a follow-up programme aimed at scaling-up community initiatives. A number of years later, this idea translated into a concrete regional project: Stimulating Community Initiatives in Sustainable Land Management.

SCI-SLM activities in Uganda started officially in 2009 with the first disbursement of UNEP-GEF's funds. However, the actual start was prior to that, thanks to an allocation of national funds. Since 2007, the Government of Uganda committed US $50,000 per annum of national resources towards the institutionalisation of the tested methodologies to further promote the role of local community initiatives in SLM. The national funds were used to conduct a number of preliminary activities (i.e. consultations with stakeholders and the identification of potential community initiatives) in preparation for the full-fledged implementation of the SCI-SLM project. After 2009, the national funds were counted as the country's 'own contribution' to the project, thus providing additional impetus to the planned activities in line with the agreed methodology for 'field activities' and 'programme development processes' (see Chapter 3).

After a brief digression on the process that led to the selection of the SCI-SLM community initiatives (CIs), this chapter provides more detailed insights on how three of the selected CIs addressed the land degradation challenge with their technical and/or social innovations. The chapter also illustrates the engagement with the communities to pilot and spread the SCI-SLM methodology.

Identification, characterisation and selection of community initiatives

The identification of potential community initiatives for SCI-SLM started in 2006 with seed funds provided by the Government of Uganda and took place in six districts of the Cattle Corridor. The districts were Ntungamo and Mbarara in the south-west, Nakasongola in the central region and Kamuli, Kumi and Soroti in the east of the country. The District Agricultural Officers (DAOs) of the districts were familiarised with the criteria to be used for the identification of the CIs. Their task was to identify three or four communities that met such criteria. In total 20 communities were identified across the six districts. A national selection team visited and screened the 20 identified communities and pre-selected the five that best met the selection criteria. The final selection took place in 2007: it was carried out in collaboration with the SCI-SLM technical advisory group. Out of the initial pool of 20 communities, four were selected to work with SCI-SLM: the Nalukonge Community Initiative Association (NACIA) in Nakasongola district, the Balimi Network for Developing Enterprises in Rural Agriculture (BANDERA) in Kamuli district, the Rwoho Environment Conservation and Protection Association (RECPA) and the Banyakabungo Twimukye Cooperative Society Limited in Ntungamo district.

The Nalukonge Community Initiative Association is a pastoral community whose land in the dryland Cattle Corridor ecosystem used to be severely affected by land degradation. Their initiative was triggered by the need to reduce the damage caused by termites and the loss of vegetative cover. NACIA was singled out for its technical innovations in the field of grazing land management including the use of arboreal termites to control terrestrial termites (through their toxic residues) and a rotational night *kraaling* system to rehabilitate degraded rangelands.

The Rwoho Environment Conservation and Protection Association distinguished itself for its unique social innovation – a community-wide movement to change the use of bare, degraded land on steep hillsides from communal livestock grazing to tree planting. Awareness raising, training, negotiations, bylaws and partnerships contributed to ensure the long-term protection of the fragile ecosystem in which the community operates.

The Balimi Network for Developing Enterprises in Rural Agriculture comprises of a large group of farmers promoting sustainable in-field practices that enhanced land productivity while, at the same time, protecting the banks of the River Nile. It stood out as a good example of combined social and technical innovation involving the setting-up, demonstration and spread of SLM practices.

Banyakabungo Twimukye Cooperative Society Limited is a common interest group focusing on livestock and grazing land management. While this initiative has been widely documented in SCI-SLM reports and factsheets, it is not included in this chapter because of its similarities with NACIA and RECPA.

The selected communities performed generally well in the TEES and/or the SRI test (see Chapter 3 for details on the tests and the selection criteria). In addition, they displayed high degrees of innovativeness and willingness to share their innovations with others. The data collected during the identification and selection phases contributed towards the progressive characterisation of the CIs and were used for the preliminary data analysis – 'stocktaking' exercise (see Chapter 11) – which took place two years after the official start of the project. Table 8.1 provides a snapshot of the characterisation for the three community initiatives presented in this chapter: NACIA, RECPA and BANDERA.

Rehabilitation of degraded rangelands: NACIA case study

Origin of NACIA and historical timeline

Following the 1986 war, the new government restructured ranches, creating smaller parcels of land for resettlement of mobile pastoralists. Each pastoralist was allocated between 40 and 60 ha of land. Nalukonge village also benefited from this scheme. However, as the livestock population increased, the originally tall and abundant grass cover diminished as a result of overgrazing and bush burning which resulted in widespread sheet erosion and surface compaction. The situation was further aggravated by charcoal burning, ostensibly to open up lands for cattle grazing. As a result, by 1998 a number of areas in Nalukonge village were seriously degraded and bare, locally referred to as '*Ebiwaramata*'. They became a common feature of the landscape. In addition, increasing destruction of the vegetation around bare areas led to termite infestation.

Table 8.1 Characterisation of SCI-SLM community initiatives (October 2011)

Categories	NACIA	RECPA	BANDERA
Location	Nalukonge village in Nabiswera sub-county – Nakasongola district	Rukoni East in Rwoho parish – Ntungamo district	Izanhiro village in Kisozi sub-county and Nalimawa village in Nawanyago sub county – Kamuli district
Organisation type	Registered association	Association officially registered with local Government	Registered limited company
Starting date	June 1998	February 2003	October 1995; but the initiative concerning the in-field SLM practices started only in 2005
No. of community members	79 Male: female ratio (67:33)	250	c. 500 Male: female ratio (70:30)
Initiative typology	Technical	Social	Social and technical
Initiative brief description	Grazing land management: experimenting with the use of arboreal termites to control terrestrial termites and the night kraaling system to rehabilitate degraded rangelands	Community Forest Management/ Ecosystem Protection: reafforestation of degraded steep slopes through a self-established environmental association	In-field SLM practices for horticulture including conservation agriculture
Origin/ source of CI	Initiated by a core group: Mr Lubega and Mr Kasasa Kezironi	Initiated by a core group: five community members led by Mr Jerome Byesigwa	Initiated by an individual: Mr George Mpaata
Motivation/ trigger of initiative	The expanding bare areas associated with termite destruction, prolonged droughts, overstocking and soil erosion	Indiscriminate harvesting of the Government's Central Forest Reserve and the harsh conditions of hot weather and prolonged drought that followed	The desire to increase the household income of group members in combination with the government's promotion of agricultural exports

Table 8.1 Continued.

Categories	NACIA	RECPA	BANDERA
Examples of CI's activities	(i) Identification of arboreal termites (AT) and assessment of their efficacy on controlling terrestrial termites (TT); cross-learning on AT ecology; introduction and monitoring the survival of AT. (ii) Construction of kraals; confining animals at night to accumulate manure; moving animals out of the kraals; reintroduction of organic matter using cattle manure to restore soil structure and nutrients to regenerate pasture; planting pasture (with stoloniferous grasses) and keeping it free of animals until established	Community awareness raising; negotiations to change land use; mobilisation of community for tree planting; controlling bush fires using local bylaws and vigorous monitoring; building partnerships to protect trees, expand the tree planting and ensure the long-term resilience of the fragile ecosystem; develop alternative livelihood activities (e.g. apiary)	Mobilisation of community to access export market through collective and sustainable farming; sensitization of the community to form an association for collective action; cross-learning on in-field SLM practices between farmers; promotion of a combination of integrated SLM practices, including construction of soil and water conservation structures, intercropping, crop rotation, diversification of crops, minimum tillage practices, composting; integrated soil nutrient management and irrigation; distribution of fruit trees and cover crop seeds; planting fruit trees in communal and individually owned gardens
Monitoring of activities	Partly	Yes	Yes
Area under SLM	c.a. 130 ha rehabilitated	80 ha of bare hills planted with trees	153 ha
Outside assistance	Yes – limited: micro-grant to buy fencing material	Yes – limited: seedlings, exchange visits	Yes – limited: some technical assistance, inputs, grants
Extra investments by community	Labour and cash	Labour and cash	Labour and cash

Table 8.1 Continued.

Categories	NACIA	RECPA	BANDERA
Problems faced by the community	Rehabilitation of highly degraded land, limited resources	Payment of subscription fee, lack of full engagement by members, limited technical skills	Land disputes, pests and diseases, declining soil fertility
Adoption/ spread: no. of communities that have adopted the CI	3	5	4
Name of adopting communities	Wadundulya, Kyabyomeire, Kyangogolo	Kafoda, Bushwere, Kagabagaba, Kagoto, Rubagano	Kasozi, Kananage Bakuseeka, Kyebajjatoboona, Butansi
Method of spread	Exchange visits	Meetings, exchange visits, training sessions, social network	Exchange visits

Source: SCI-SLM monitoring data.

In 1998, two community members, Mr G.W. Lubega and Mr Kasasa Kezironi, observed the expansion of bare areas on their land and requested the Migeera Local Council II chairperson, Mr Paul Mugume, to convene a community meeting to discuss the challenge at hand. The meeting took stock of the scale of the problem and it was agreed to establish an association for the rehabilitation of the degraded rangelands. The Nalukonge Community Initiative Association (NACIA) was founded. In follow-up meetings, the community developed an action plan. Construction of valley tanks in each of the six former ranch zones was identified as a way of reducing crowding of livestock around water sources which, in turn, caused overgrazing. Second, the association identified the need to control sheet erosion on degraded grazing land. With the technical support of the Ministry of Agriculture, Animal Industry and Fisheries (MAAIF) and the local government, work was undertaken to rehabilitate the degraded areas – tree planting and erosion control measures (e.g. contour bunds stabilised with grass) were used. However, termites destroyed the planted trees and grass, thus limiting the level of success. The large extent of the areas affected made it unviable and environmentally unsustainable to control termites using chemicals. At that critical moment, the ingenuity and the proactive nature of the community became instrumental in breaking the vicious cycle of degradation – through innovation.

One of the local farmers noticed that grass grew particularly well where animals deposited their dung. The effect was even more visible in areas where night *kraals*[5] were in use. Therefore, rotational night *kraaling* was chosen by the community members as a valuable option to experiment with. However, the solution to the multifaceted problem experienced by NACIA required a basket of remedies. During one of his trips, Mr Lubega had learnt that the communities in Kamwenge and Kiruhura districts in south-western Uganda use the arboreal termite species (*Microcerotermes edentatus*) that nest in trees to control the spread of terrestrial termites. The idea was then put forward for further investigation.

It was around that time that the collaboration between MAAIF and NACIA developed, bringing into the picture new partners (the National Agriculture Research Organisation, NARO, and Makerere University) and projects (UNDP/Small Grants Programme and SCI-SLM) interested in working with the community and helping it improve its own initiative. The collaboration between NACIA and these stakeholders played an important role in fostering the joint experimentation of the rotational *kraaling* system for the rehabilitation of degraded rangelands and the use of arboreal termites to control terrestrial termites. Over 10 years, and with some external support, it was possible for NACIA to turn the initial ideas and observations into a sustainable solution to its environmental and socio-economic problem.

The timeline of key events that led to the creation and the development of NACIA as well as the list of external actors that supported NACIA in the improvement of its initiative are presented in Table 8.2.

Description of the community initiative

NACIA was selected as a SCI-SLM community for its distinctive use of indigenous knowledge and local innovation in SLM to address land degradation. In line with the SCI-SLM approach, NACIA was keen on sharing its knowledge and experience with other communities. It was also driven by the ambition of further improving its initiative by working in partnership with Makerere University, the National Agriculture Research Organisation and other relevant actors. NACIA was singled out for its technical innovations, with high adoption potential, in the field of grazing land management.

(i)   Night kraaling to rehabilitate degraded rangelands

The technique involved the construction of night *kraals* on degraded bare surfaces by using branches of invasive woody species as a temporary fence. Cattle were confined in temporary *kraals* at night until a layer of approximately 5 cm of animal manure accumulated on the soil surface. This was done during the dry season. At the onset of the rains, to supplement the naturally spreading stoloniferous grasses (stimulated by the manure), appropriate grass and legume seed species were broadcast on the manure. The *kraal* was then moved to another denuded site. The original *kraal* area was kept free of animals until pastures were established and started producing seed to restore the soil seed bank. Two months after the establishment of pasture seedlings, calves were allowed to graze the established pasture to enhance development of tillers.

This technique improved degraded bare surfaces through accumulation of cattle manure (Figure 8.2). It enabled grasses to germinate, thus restoring the vegetative cover in the grazing land (Figure 8.3). The grass cover depended on the quantity of accumulated manure. The technique improved pasture dry matter production to a

Table 8.2 Historical timeline of NACIA

Year	Activity	External actors
1986	End of the war in the region	
1989	Restructuring of the ranching schemes	Government
1994	First signs of serious degradation observed in Migeera parish	Local pastoralists and ranch owners
1998	Formation of NACIA to address rangeland degradation	Community members
1999	Initial support to halt erosion; construction of valley tanks to harvest water and tree planting in the bare areas	MAAIF and UNSO-UNDP
2002	Formal registration of NACIA as an association	NARO
2003	Joint experimentation with Makerere University on soil erosion control and night *kraaling* to re-vegetate bare areas	Makerere University
2004	Study tour of NACIA members to Mbarara and Ibanda districts in south-western Uganda to learn about water harvesting using waterproof tarpaulin and biogas	MAAIF and Intergovernmental Authority on Development (IGAD)
2006	MSc students of the Vrije Universiteit Amsterdam documented NACIA innovations on night kraaling	Vrije Universiteit Amsterdam and SCI-SLM
2006–2007	NACIA was identified and selected as a SCI-SLM community initiative	District Agricultural Officers, MAAIF, SCI-SLM Uganda and TAG
2007	First attempt to introduce arboreal termites from the south-western Uganda district of Kamwenge to Nakasongola	MAAIF and SCI-SLM
2008	Laboratory experimentation and initial field release of arboreal termites to establish their effects on terrestrial termites	NARO and MAAIF
2011	Identification and taxonomic and ecological classification of arboreal termites	NARO and Royal Museum of London
2011	Exchange visit of BANDERA representatives to NACIA	SCI-SLM and BANDERA
2012	Dissemination of results of experimentation on the impact of arboreal termites on terrestrial termites	NARO and MAAIF
2012	NACIA Chairperson participates in the Regional Steering Committee meeting and cross-learning visits hosted in Ghana	SCI-SLM
2012	Study tour to Kamwenge district on the use of arboreal termites to control terrestrial termites	NARO, MAAIF and SCI-SLM

Table 8.2 Continued.

Year	Activity	External actors
2012	Introduction and field distribution of arboreal termites from Kamwenge to Nakasongola	NARO and MAAIF
2012	Planning meeting to upscale rehabilitation of degraded rangelands using the night kraaling and arboreal termites	NARO, MAAIF and SCI-SLM
2013	Monitoring visits to assess field establishment of the arboreal termites in Nakasongola	NARO, MAAIF and SCI-SLM

Source: SCI-SLM monitoring data.

local maximum of 4,500 kg/ha. The increase in dry matter production led to dramatic improvement of the area's carrying capacity. The technique was also associated with the improvement of soil pH from 3.4 per cent to 5.8 per cent, organic matter from 1.3 per cent to 3.1 per cent and nitrogen from 0.07 per cent to 0.20 per cent.

(ii)   Use of arboreal termites to control terrestrial termites

The joint experimentation started with laboratory identification of the arboreal termite species (*Microcerotermes edentatus*) at the National Agriculture Research Laboratories Institute (NARLI). Laboratory investigations confirmed over 80 per

Figure 8.2 Manure accumulated as a result of night *kraaling*
Source: W. Critchley 2005.

Figure 8.3 Grass in the process of re-establishment on bare land
Source: W. Critchley 2005.

cent mortality in terrestrial termites within 48 hours of the introduction of arboreal termites. Dead specimens changed colour from brown to black, showing the effect of arboreal termites on terrestrial termites. Previous studies demonstrated that the deposits of termites produce a number volatile fatty acids with repellent properties (McFarlane 1984), while others cause mortality of insects (Adebowale and Adedire 2006). The controlling effect of the arboreal termites may therefore be associated with the repelling effects of volatile toxic chemicals.

After the experiments in the laboratory, the field sites for the release of the arboreal termites were jointly identified by the researchers and the community members. The release sites fulfilled the following conditions: (i) the availability of a mature tree with a broad canopy; (ii) a cool environment; and (iii) a source of decomposing plant material under the tree. With time, the arboreal termites construct tunnels up the tree trunk, which aid their movement to feed on decomposing organic matter on the ground. Appearance of nests on tree branches after 6–9 months is a sign of establishment of the arboreal termites. The time of encounter with the destructive terrestrial termites is during feeding, when their presence on the ground is thought to deter the terrestrial termites in the environment.

The arboreal termites were found in only six out of 17 release sites after seven months. The relatively higher temperatures and the longer dry periods that prevail in Nakasongola from December to March compared to Kamwenge is the most likely factor responsible for the low survival of the arboreal termites. However, the communities are confident of higher chances of establishment at the recovery sites, having overcome the stress period. The community members monitor the sites with the expectation that these will be the source material for scaling-out the technique to other areas within Nalukonge. Should the arboreal termite technology be found to be effective in the long run, it would be a cheaper and more sustainable alternative to insecticides as it involves small investments in field multiplication, mainly by planting suitable tree species for nesting.

Added value of working with SCI-SLM

SCI-SLM recognised the value of indigenous knowledge and local innovation in providing sustainable solutions to the pernicious problem of land degradation. It stimulated the community initiatives and encouraged partnerships with research and academic institutions to bridge knowledge gaps but also create new knowledge – 'hybrid knowledge' – especially through joint experimentation. The project provided methodological guidance for the identification, selection, characterisation and validation of NACIA's technical innovations. Attention was equally given to the social aspects of the community initiative even when they did not qualify as innovations. SCI-SLM strengthened community planning and monitoring. Furthermore, it enhanced the capacity of the community to identify strategic partners who were able to support the improvement and horizontal spread of the initiative. SCI-SLM facilitated cross-learning at various levels from local to regional. Farmer-to-farmer learning visits, in particular, exposed NACIA's community members to the knowledge, practices and experiences of communities from other parts of Uganda and beyond (e.g. termite control using arboreal termites in Kamwenge district and conservation agriculture practices from BANDERA). Similarly, NACIA provided a learning platform for other communities. For instance, as a result of the exchange visit to NACIA in September 2011, members of the BANDERA community appreciated the efforts made to fight land degradation in Nalukonge and, with the technical guidance of NACIA, adopted the construction of underground water harvesting tanks.

Adoption of the CI: towards horizontal spread

After just two years following the involvement of NACIA in the SCI-SLM project, their initiative for the rehabilitation of degraded rangeland had spread to another three communities (with no identified innovations), Wadundulya, Kyabyomeire and Kyangogolo, thanks to the exposure visits organised at district level. Horizontal spread also occurred as a result of the exchange visits between innovative communities working with SCI-SLM. For example, following a visit to NACIA in 2011, the BANDERA community wanted to learn how to construct the type of underground tarpaulin-lined water harvesting tank that they had seen in Nalukonge. Under the framework of the SCI-SLM project, a member of NACIA, Mr Kakungulu, trained the BANDERA community. Since then, BANDERA members have constructed 16 such water harvesting tanks on their own.

In order to achieve further horizontal spread of the initiative, NACIA developed a proposal to rehabilitate 10 more degraded rangeland sites in the Nalukonge community. The proposal was prepared in collaboration with NARO, MAAIF,

NEMA, Makerere University and the District Local Government. It was then submitted to the UNDP/MAAIF SLM Enabling Environment Project for funding. SCI-SLM was instrumental in organising and facilitating the planning meeting that led to the development of the small grant proposal.

Lessons learnt

Two main lessons were drawn from the experience with NACIA: (i) the need for continuous monitoring of initiatives; and (ii) the value of indigenous knowledge. It was noted that rigorous and long-term monitoring of initiatives is necessary not only for their validation, but also to support evidence-based adoption (e.g. establishment, survival and effect on terrestrial termites to ascertain the impacts of arboreal termites). As for indigenous knowledge, it is acknowledged by many as a key factor in bridging the knowledge gap and shaping local community innovations – but its potential has not yet been exploited.

Highlands' ecosystem protection with community forest management: RECPA case study

Origin of RECPA and historical timeline

The founding of the Rwoho Environment Conservation and Protection Association (RECPA) was the answer of the local community to the challenging environmental conditions that followed the harvesting through clear felling of the Government's Central Forest Reserve (CFR) in Rwoho parish in 2002. The residents felt that, since the clearing of the forest, the area had been affected by unprecedented hot weather and a severe drought. Firewood had also become increasingly scarce. The unbearable living conditions experienced for several months prompted some members of the community to take their own initiative to mitigate the negative impacts of deforestation. It was as result of the collective will and action of the community members that RECPA was officially established in February 2003.

Over the years RECPA has developed into a social initiative with considerable potential – a community-wide movement to change the use of bare, degraded land on steep hillsides from communal livestock grazing to tree planting. As stated in its founding documents, the aims of RECPA are to (i) establish continuous and long-term soil protection on steep and bare slopes through tree planting; (ii) control bush fires using local bylaws and monitoring; and (iii) promote alternative livelihood activities (e.g. beekeeping). To achieve the desired ecosystem protection, the initiative has focused on influencing land use change on public and privately owned degraded steep hills, shifting from communal livestock grazing and cropping to individually owned and managed plantation forests. This initiative has contributed to secure land tenure and encouraged long-term investment in tree planting. Furthermore, it has generated multiple environmental benefits in terms of soil protection, flood prevention, improved microclimate and a more reliable provision of ecosystem services, thus safeguarding the (economic) investments made by the community.

An important catalyst for the RECPA initiative has been the implementation of the Rwoho CFR carbon project under collaborative forest management (CFM), with the National Forestry Authority (NFA). The CFM project (started in 2006 under the auspices of the World Bank and the European Union) incentivises the community to plant trees for carbon sequestration purposes. RECPA was offered 60 ha in the Rwoho Central Forest Reserve to implement the project. By January 2013, nearly 40 ha had been planted and the benefits accruing to RECPA members under the carbon-trading mechanism were estimated to be equivalent to US $4,602. By 2017, which is the term set for the second verification and payment of the carbon credits, RECPA members hope to have the entire area of 60 ha completely afforested.

The timeline of key events that led to the creation and the development of RECPA as well as the list of external actors that supported RECPA in the improvement of its initiative are presented in Table 8.3.

Description of the community initiative

The RECPA initiative is a social innovation based on the mobilisation and organisation of community members to promote land use changes that reduce environmental hazards, ensure land restoration and increase ecosystem services. It entails a broad-based awareness campaign and continuous dialogue with landowners, local leaders and other natural resource users to work towards sustainable use and management of the fragile highlands' ecosystem. The dialogue with local leaders and land users (particularly herders and farmers) is combined with the enforcement of measures aimed at ensuring the protection of forests against fires and livestock overgrazing. These measures include penalties for offenders (e.g. fines, obligation to replant the forest and even imprisonment) as well as incentives for those denouncing the culprit.

The initiative is carried out on individually owned land and public land allocated under a long-term agreement with the National Forestry Authority. While RECPA provides a common framework for the activities of the community members, the actual implementation, investments and benefits are realised at household level. The tree-planting technique practised by RECPA has no innovative aspects. It is a conventional practice already used in the plantation systems of the Rwoho Central Forest Reserve. At the onset, the initiative relied on the tree-planting skills (e.g. setting up tree nurseries) of the few RECPA members who had previously worked in the CFR. Over time, these skills have been gradually passed on to other community members.

Since its inception in 2003, RECPA has afforested a considerable area of previous bare hills in Rwoho parish. Approximately 50 ha of private land and nearly 45 ha of public land under the CFR have been rehabilitated with tree planting. RECPA members have also established ten private tree nurseries in Rwoho parish. Even though livestock rearing has been negatively affected by the establishment of tree plantations, RECPA members recognise that the increase in tree and vegetation cover has had a positive impact on the overall productivity of the highlands. However, the plantations consist mostly of pine and eucalyptus trees. This is a choice made by the community in consideration of their resilience, high economic value

Table 8.3 Historical timeline of RECPA

Year	Activity	External actors
2002	The government's Central Forest Reserve (CFR) in Rwoho parish was harvested by the National Forestry Authority (NFA)	
2002–2003	Rwoho parish experienced harsh climatic conditions and a severe drought	
2003	First brainstorming meeting of Rwoho parish residents (9 people) to find solutions to the problem; residents resolved to plant trees; they agreed to form an association: RECPA	Rwoho parish residents led by Mr Jerome Byesigwa
2003	RECPA members who previously worked in the CFR started establishing privately owned tree nurseries	
2004	Two MSc students of the Vrije Universiteit Amsterdam documented the RECPA initiative	Vrije Universiteit Amsterdam and SCI-SLM
2004–2008	Planning, training, awareness raising, bylaw formulation and negotiations to acquire and utilise land from the Central Forest Reserve in Rwoho	National Forestry Authority and National Environment Management Authority (NEMA)
2006–2007	RECPA was identified and selected as a SCI-SLM community initiative	District Agricultural Officers, MAAIF, SCI-SLM Uganda and TAG
2006	Further MSc students of the Vrije Universiteit Amsterdam studied the initiative	Vrije Universiteit Amsterdam and SCI-SLM
2006	RECPA engaged in a carbon project under collaborative forest management	NFA, European Union, World Bank
2008–2009	MSc students of the Vrije Universiteit Amsterdam followed up and analysed the initiative	Vrije Universiteit Amsterdam and SCI-SLM
2011	Characterisation and validation of the RECPA initiative	MAAIF and SCI-SLM
2011	A fourth pair of MSc students of the Vrije Universiteit Amsterdam further studied specific aspects of the RECPA initiative	Vrije Universiteit Amsterdam and SCI-SLM
2011	Learning visits to Banyakabungo community in Ntungamo district and other groups in Kabale district	Banyakabungo community, SCI-SLM and MAAIF
2011	RECPA's chairperson participated in a skill-sharpening workshop facilitated by SCI-SLM in Kampala	MAAIF, SCI-SLM Uganda and TAG
2011	As part of the SCI-SLM third Regional Steering Committee meeting held in Uganda, RECPA hosted an exchange visit with representatives of SCI-SLM community initiatives from Ghana, Morocco, South Africa and Uganda	SCI-SLM and community representatives from Ghana, Morocco, South Africa and Uganda

Table 8.3 Continued.

Year	Activity	External actors
2013	50 ha of private land and nearly 40 ha in the forest reserve rehabilitated with tree planting	World Bank
2013	RECPA community and their agriculture extension officer trained on the use of soil-testing kits	SCI-SLM and National Agricultural Advisory Services (NAADS)

Source: SCI-SLM monitoring data.

and quick maturation compared to indigenous species, but not the most appropriate choice from an ecological/environmental point of view. This important technical point has been taken on board by RECPA since its involvement in SCI-SLM.

The achieved results in terms of ecosystem protection came with a further challenge. With most land on hilltops and steep slopes taken up by tree plantations and livestock grazing claiming part of the farming land, it became necessary to intensify crop production in order to meet the immediate food and income needs of RECPA members. In order to achieve a sustainable intensification of crop production, RECPA engaged with several organisations that could provide training, demonstration materials and suitable equipment. For instance, the National Agriculture Advisory Services (NAADS) district office supported RECPA with training and the provision of high-yielding varieties of banana, potatoes, beans and maize, while Makerere University provided training on soil fertility management and the use of a soil-testing kit.

Added value of working with SCI-SLM

The SCI-SLM project helped RECPA recognise the value and potential of its own initiative but also understand how it could be improved. Thanks to SCI-SLM, RECPA could better appreciate the link between their social innovation and the technical SLM aspects. For example, SCI-SLM encouraged RECPA to diversify the tree planting by including indigenous trees, especially in the lower valleys where they are better adapted. It also identified poor soil fertility management as a barrier to increased production. Therefore, in partnership with MAAIF and Makerere University College of Agricultural and Environmental Sciences, SCI-SLM supported RECPA with soil-testing kits, relevant training and soil sample analysis. It also promoted sustainable intensification of agricultural activities. Furthermore, SCI-SLM advised RECPA to complement tree planting with other livelihood activities (e.g. beekeeping) and alternative (non-timber) use of plantation forests (e.g. controlled grazing).

The involvement in SCI-SLM also helped RECPA to realise possible short-comings of its social innovation. By 2011, RECPA numbered 230 registered members (of which 195 were men and 35 were women; only 10 among them were young adults). The gender imbalance was explained by the fact that initially only landowners were granted membership. The advisory role of SCI-SLM was instrumental in ameliorating the situation. In the same year, RECPA adopted a resolution that allowed every woman married to a member of the association to become automatically a member with equal rights and responsibilities.

In addition, SCI-SLM supported the documentation and the spreading of the community initiative. With the facilitation of MAAIF, a total of nine Master's students of the Vrije Universiteit Amsterdam – under supervision of SCI-SLM's technical advisory group – spent a considerable length of time living in the community to analyse and document specific aspects of best practices and innovations jointly with RECPA members. The MSc theses produced contributed to raise attention on the initiative both nationally and regionally. The residential visits by students were matched with backstopping visits of the TAG and the national coordination team. These visits, which were planned for the identification, selection and characterisation of the initiative, had the added value of further encouraging and motivating the community.

RECPA also benefited from the exchange visits to other SCI-SLM communities, especially the Banyakabungo community in Ntungamo district and other groups in Kabale district. Finally, as part of the SCI-SLM third Regional Steering Committee held in Uganda in October 2011, RECPA proudly hosted an exchange visit with representatives of the SCI-SLM communities from Ghana, Morocco, South Africa and Uganda.

Adoption of the CI: towards horizontal spread

As part of the project, the SCI-SLM national coordination team trained the National Agriculture Advisory Services providers in Ntungamo district on the SCI-SLM methodology with the intention of spreading the approach to other communities. Two years after the official involvement of RECPA in the SCI-SLM project, their initiative had already spread to another five communities (with no identified innovations), Kafoda, Bushwere, Kagabagaba, Kagoto and Rubagano, thanks to the exposure visits, meetings and training sessions organised at district level. The social network of local contacts also played a role in this respect. Under RECPA's inspiration, these five communities were encouraged to join the Rwoho CFR Carbon Financing project.

While RECPA's social innovation does not present significant barriers to horizontal spread, the associated tree planting may be a constraint – at least in the short term. In fact, in the long run the high returns on tree plantation investments more than compensate for the cost of bringing timber trees to maturity (approximately 15 years), thus providing an incentive to households with substantial land area to invest in tree plantations. In the medium term, some economic benefits can also be achieved from the opportunistic harvesting of trees (e.g. during thinning), the integration of forest-friendly activities (e.g. beekeeping), controlled grazing and harvesting of firewood by the community. However, the lack of short-term economic benefits, the high cost of establishment and maintenance of the tree plantations and the continuous risk of fire may pose a limit to a wider spread of the initiative.

Lessons learnt

Two main factors contributed to RECPA's successful experience: (i) security of tenure over land and natural resources; and (ii) access to financial incentives. It was noted that the secure land tenure systems in Rwoho parish gave the necessary confidence to the land users to make long-term investments on the land (e.g. tree planting). This worked well in combination with the financial incentives provided by the carbon-financing project, which are over and above the other benefits provided by the planted trees (e.g. ecosystem goods and services).

In-field sustainable land management practices: BANDERA case study

Origin of BANDERA and historical timeline

The Balimi Network for Developing Enterprises in Rural Agriculture is a community initiative driven by the idea that access to secure markets is a pre-condition to SLM. Their assumption is that a reliable source of income from farming encourages people to invest back into the land, thus strengthening the long-term resilience of the land and the household.

The initiative was triggered by a number of factors. Since the 1990s, the livelihoods of people in Kisozi and Nawanyago sub-counties in Kamuli district were threatened because of the rapid increase in population, which resulted in the reduction of the average landholding to 1 ha per household and a decline in the yields of staple crops. The increased competition for land between food crops and industrial crops (e.g. sugar cane) further exacerbated poverty and household food insecurity. These factors prompted a number of community members, led by Mr George Mpaata, to explore innovative ways of increasing their incomes, also taking advantage of the government's drive for agricultural exports. This process led to the formation of the BANDERA community in 1995. It started as a silk production enterprise with a guaranteed export market in Japan. The membership further expanded with diversification to other export-driven enterprises, particularly okra, chillies and spices (1997–2004) as well as fruit and horticulture production (2005–2008).

In 2005, confident in their ability and motivated by the export companies that they were working with, the group rented land in Kiyunga parish along the banks of the River Nile. They established a nucleus farm of 153 ha, managed by 80 members (45 women and 35 men), producing mangoes, citrus, passion fruits and pineapples under irrigation. In addition to the production from the nucleus farm, a number of outgrowers were enlisted by the group from the neighbouring 11 sub-counties. BANDERA was successful in using its strong social organisation skills to mobilise a sizeable community of small-scale producers and motivate them to farm collectively on a commercial scale to benefit from the promotion of agricultural exports by the Government of Uganda. The community was also successful in matching its business ambition with the use of SLM practices to protect the banks of the River Nile. These were key factors in the selection of BANDERA as a SCI-SLM community in 2007. This initiative inspired many and attracted the attention of civic and political leaders. Among others, BANDERA was visited by H.E. the Vice President Prof. Gilbert Bukenya (October 2006) and H.E. the President Yoweri Kaguta Museveni (May 2007).

Despite the remarkable achievements and the heavy investments made by the group on the rented nucleus farm, in October 2008 the community lost the entire area under cultivation due to a change in land ownership. Refusing to give up hope, in 2009, BANDERA relocated its head office to Nalimawa village in Nawanyago sub-county. Building on lessons learnt[6] and with the advice of SCI-SLM, BANDERA changed its approach and supported its members with the integration of SLM practices into individually owned farm holdings. Since then, BANDERA has grown into a model community group with an established learning centre for sustainable land management in smallholder farming enterprises – with a current focus on small-scale conservation agriculture.

The timeline of key events that led to the creation and the development of BANDERA as well as the list of external actors that supported BANDERA in the improvement of its initiative are presented in Table 8.4.

Description of the community initiative

BANDERA comprises a large group of farmers promoting sustainable in-field practices that enhance land productivity while, at the same time, protecting the environment. This initiative stood out as a good example of a combined social and technical innovation involving the setting-up, demonstration and spread of SLM practices. While the social innovation relates to the approaches and strategies of community organisation to enhance the adoption and horizontal spread of the initiative, the technical innovation involves the integrated use of in-field SLM practices (e.g. conservation agriculture, soil fertility management, water harvesting) to increase productivity.

Below are the principle examples of in-field SLM practices used by the BANDERA community.

Soil and water conservation/water harvesting structures

In individually owned farms, water retention channels are dug along contours to conserve runoff water and soil. Usually, napier grass (*Pennisetum purpureum*) is planted because of its multiple uses – it protects the soil bunds but it can be also used for mulching and as a livestock fodder. Other in-field soil and water conservation structures include: the use of grass strips and trash lines along contours to prevent soil and water erosion; the use of large basins around tree crops to collect and concentrate rainfall runoff through water harvesting; channels connected to farmland to harvest road runoff.

Conservation agriculture

In 2011, BANDERA's chairperson participated in a skill-sharpening workshop facilitated by SCI-SLM in Kampala. It was during this event that he developed an

Table 8.4 Historical timeline of BANDERA

Year	Activity	External actors
1995	BANDERA was formed by a group of silkworm farmers led by Mr George Mpaata in Kiyunga Village, Kamuli District	
2005	BANDERA set up a 153-ha nucleus farm managed by 80 farmers (45 women and 35 men). The total membership was 501 including outgrowers	
2006–2007	BANDERA was identified and selected as a SCI-SLM community initiative	DAO, MAAIF, SCI-SLM Uganda and TAG
2007	BANDERA was visited by H.E. the President of Uganda Yoweri Museveni	H.E. The President of Uganda
2008	BANDERA was evicted from the nucleus farm	Land owner
2009	BANDERA established its new head office in Nalimawa village	
2011	SCI-SLM held a planning meeting with BANDERA	MAAIF and SCI-SLM
2011	Study tour to Iganga, Pallisa and Soroti to learn about fruit tree management and agroforestry	MAAIF and SCI-SLM
2011	Two MSc students of the Vrije Universiteit Amsterdam studied BANDERA's social and technical innovations	Vrije Universiteit Amsterdam and SCI-SLM
2011	BANDERA's chairperson participated in a skill-sharpening workshop facilitated by SCI-SLM in Kampala and expressed interest in conservation agriculture (CA)	MAAIF, SCI-SLM Uganda and TAG
2011	Established demonstration plot on CA in Nalimawa village	MAAIF and SCI-SLM
2011	Exchange visit of NACIA representatives to BANDERA	SCI-SLM and NACIA
2011	As part of the SCI-SLM third Regional Steering Committee meeting held in Uganda, BANDERA hosted an exchange visit with representatives of the SCI-SLM community initiatives from Ghana, Morocco, South Africa and Uganda	SCI-SLM and community representatives from Ghana, Morocco, South Africa and Uganda
2011	Representatives visit RECPA and Banyakabungo	SCI-SLM
2011	BANDERA trained in the construction of underground water harvesting tanks by a member of NACIA	NACIA, SCI-SLM and MAAIF

Table 8.4 Continued.

Year	Activity	External actors
2011	Study tour to Pallisa to learn about conservation agriculture	MAAIF and SCI-SLM
2012	The community hosted a Training of Trainers on conservation agriculture for members from six districts in the Cattle Corridor	MAAIF, SCI-SLM and Rural Enterprise Development Services (REDS)
2012	UNDP Country Director and SLM Project Board members visited BANDERA	MAAIF and UNDP
2012	BANDERA set up 30 CA demonstration plots in Nawanyago and Kisozi sub-counties	MAAIF and SCI-SLM
2012	An MSc student of the Vrije Universiteit Amsterdam studied the BANDERA initiative in conservation agriculture	Vrije Universiteit Amsterdam and SCI-SLM
2012	The community hosted the national celebrations for the World Day to Combat Desertification (WDCD)	MAAIF
2012	BANDERA exhibited a CA demonstration at the Annual National Agricultural Show held in Jinja district, Uganda	MAAIF, SCI-SLM and Common Market for Eastern and Southern Africa (COMESA)
2012	The chairperson participated in the SCI-SLM fourth Regional Steering Committee meeting in Tamale, Ghana	SCI-SLM
2013	BANDERA's Model Farmer Resource Centre launched by the Speaker of the National Parliament	MAAIF

Source: SCI-SLM monitoring data.

interest in conservation agriculture. BANDERA decided to apply CA's three principles – (i) minimum tillage (by constructing permanent planting basins in this case[7]); (ii) crop rotation, including legumes; (iii) continuous soil cover – and assess their impact on the productivity of the land. With the assistance of MAAIF and SCI-SLM, the community set up a small CA experimental plot (0.25 ha) where different combinations were tested on maize and beans (in planting basins: DAP phosphate fertilizer and urea; decomposed cow manure; no fertilizer or manure; or using the conventional planting method without basins and with no fertilizer or manure applied).

Table 8.5 presents the comparative inputs and measures used for the testing of conservation agriculture and conventional farming practices in BANDERA's experimental plot in 2012.

In the season April–July 2012, the average yield for maize under the conventional system was 1.25 tons (t)/ha compared to 4.6 t/ha under CA, while the average yield for beans during the same season was 0.32 t/ha under the conventional system

Table 8.5 BANDERA joint experimentation on CA and conventional practices: inputs and measures

	Maize		Beans	
	CA practices	Conventional practices	CA practices	Conventional practices
Time of land preparation	After previous harvest	At onset of rains	After previous harvest	At onset of rains
Tillage	Minimum tillage	100% of the land tilled	Minimum tillage	100% of the land tilled
Planting stations	Permanent planting basins (15cm W x 35cm L x 15cm D)	Conventional planting holes (3–5cm deep)	Permanent planting basins (15cm W x 35cm L x 15cm D)	Conventional planting holes (3–5cm deep)
Spacing	35cm within row; 75cm between rows	60cm between holes in a row; 90cm between rows	35cm x 75cm	Staggered planting, average 25cm between planting stations
No. of plants per station	3	2	6–8	2
Fertilizers	125kg diammonium phosphate (DAP)/ ha at planting; 125kg urea/ha at 3 weeks	None	DAP (occasional)	None
Weed control	Glyphosate, mulch, light surface weeding	Hand-hoe weeding	Glyphosate, mulch, light surface weeding	Hand-hoe weeding
Crop residue management	Used to cover the soil as mulch	Fed to livestock, burnt, fuel or occasionally left in the garden	Used to cover the soil as mulch	Fed to livestock, burnt, fuel or occasionally left in the garden

Source: SCI-SLM monitoring data 2012 (MAAIF).

and 0.93 t/ha under CA. Farmers attributed the increase in yields to the following factors: (i) early planting; (ii) increased storage of moisture in the permanent planting basins; (iii) higher plant population under the precise basin spacing; and (iv) addition of soil fertility-enhancing inputs (fertilizers, manure and mulch). These encouraging results inspired farmers to increase the acreage under CA for these crops in the following season. It also motivated them to plan their activities judiciously to improve their production methods. However, before upscaling, they decided to make good use of their participation in SCI-SLM to gain more exposure and better understanding of CA practices.

Added value of working with SCI-SLM

BANDERA benefitted from the collaboration with SCI-SLM on many accounts.

Improvement of the community initiative

Participation of the community in the characterisation and validation of the innovations using the SCI-SLM methodology (TEES and SRI tests) strengthened their understanding of the link and contribution of their innovations to SLM. Also, the regular visits to the community by the SCI-SLM national team and the TAG as well as the District Agricultural Officers motivated the group to improve and spread their initiative. SCI-SLM also promoted the use of appropriate tools to support land management choices (e.g. field soil-testing kits).

Study tours and exchange visits

SCI-SLM facilitated several farmer-to-farmer learning activities (e.g. open days) which provided a good opportunity for sharing relevant technology and know-how between communities in a simple, practical and cost-effective manner. It also supported the study tour to Pallisa of ten BANDERA members to learn about conservation agriculture.

Joint experimentation

The SCI-SLM methodology and approach which encourages joint experimentation, building on indigenous knowledge, provided an effective platform for the interaction of BANDERA with relevant research institutions, particularly the National Agriculture Research Laboratories Institute.

Partnership building

SCI-SLM was instrumental in forging the partnership of BANDERA with REDS, MAAIF and UNDP. This partnership resulted in more support for the community initiative through input provision and technical training in CA. With assistance of the SLM Mainstreaming project managed by MAAIF, it was possible to establish the CA demo plots. In particular, 30 members of the BANDERA community were each provided with inputs to set up a 0.2 ha demonstration plot of maize and equally sized plots of beans to further experiment on the technology during the first rainy season (April–July 2012). While farmers provided labour, land and other locally available inputs (e.g. string rolls, pens, protective gear and record books), MAAIF provided the external inputs needed for the joint experiment (e.g. fertilisers, herbicides, seeds and spray pumps).

Monitoring and documentation
The student exchange programme between the SCI-SLM project and the Vrije Universiteit Amsterdam provided an additional boost to the research and documentation of the community initiatives.

Adoption of the CI: towards horizontal spread

BANDERA used several strategies to promote and spread its initiative. The focus was predominantly on conservation agriculture. Below are some examples of the alternative approaches used.

- *Mobilisation of existing social networks* – Members of the BANDERA community involved other interested community members, including their neighbours, friends, relatives and local associations in setting up and monitoring the CA demo plots. Through this effort, at least 27 of the group members who established demonstration plots reached out to a minimum of three other community members who took up the practice. For instance, Mr. Ali Kasadha from Kiyunga village trained 52 farmers from the neighbouring Kayunga district, while Mr. Jamada Isabirye from the same village reached out to 20 farmers in his parish.
- *BANDERA outreach and training team* – BANDERA's promising results with CA created increasing demand from local leaders to introduce the technology to other farmers. To meet this demand, BANDERA established an outreach and training team consisting of five members (3 men and 2 women). The team targeted other interested community-based organisations (CBOs) and local leaders, as well as medium- and large-scale farmers within a radius of 30 km, offering practical training in various CA practices. By the end of 2013, the team had reached out to 315 farmers after carrying out 31 training sessions and facilitated the establishment of 108 CA gardens.
- *BANDERA Model Farmer Resource Centre* – As a result of the high publicity of the BANDERA community initiative and the high frequency of visits from farmers, the community designated their main technology demonstration farm as a Model Farmer Resource Centre. The centre opened in July 2013. It offers demonstrations of various land management practices for annual and perennial crops and livestock production. These include water harvesting, soil and water conservation, conservation agriculture, integrated nutrient management, zero grazing, biogas production, drip irrigation, agroforestry, integration of crops and fruit and yogurt processing. Since June 2012, a total of 33 farmer and other interest groups have visited the resource centre. The Kamuli District Local Government Executive Committee was also on the list of visitors.
- *Farmer-to-farmer visits and open days* – 17 members of the BANDERA community visited Soroti and Pallisa districts to learn about the management of citrus orchards. Some 14 members of the NACIA community in Nakasongola district visited BANDERA. As a result of these visits, five BANDERA members doubled their areas under citrus orchards and 21 farmers improved in-field SLM practices in their orchards, including water retention basins, organic manure application and mulching. Similarly, 10 BANDERA members were trained in constructing

underground water tanks by a NACIA member and have since constructed 14 similar water tanks in the community for household use, drip irrigation and watering livestock. Through regular open days, the BANDERA community has been visited by numerous farmer groups with the facilitation of the National Agriculture Advisory Services. In addition, other SLM projects have used BANDERA's in-field SLM demonstration centre to train their farmers. For example, 180 farmers from 14 groups under the SLM mainstreaming project in six neighbouring districts visited BANDERA community, particularly to learn about CA practices.

- *Public awareness* – The high maize and bean crop yields under CA encouraged BANDERA and other beneficiary communities to share their experience on radio and television. Since 2012, BANDERA members have featured more than 15 times on local radio programmes. In addition, at least seven newspaper articles featured BANDERA community activities in 2012. The BANDERA community hosted the national activities to mark the World Day to Combat Desertification (17 June 2012) under the theme, 'Healthy soil sustains your health: Let's go land-degradation neutral'. The three-day event attracted over 3,000 participants who were shown exhibitions and demonstrations on SLM practices.

Lessons learnt

The success of the BANDERA community initiative provided motivation for adoption of the SCI-SLM approach and methodology by other SLM projects and initiatives in Uganda. The winning factor is that the community is at the centre of the SCI-SLM approach and all activities (e.g. characterisation, validation, cross-learning) are meant to stimulate and maximise its innovative potential. In the case of BANDERA, the participation in SCI-SLM acted as a springboard. Not only did the community become confident about its initiative through its increased knowledge of SLM but it also became eager to positively influence others. Another important lesson concerns the strategic forging of partnerships between the community, researchers and academics, which opened up unforeseen opportunities for joint experimentation to explore new ways of dealing with land degradation and productivity challenges.

Conclusions

The experiences of NACIA, RECPA and BANDERA provide field evidence to support a number of insights:

- Indigenous knowledge of rural communities is a critical factor in the quest for local and sustainable solutions to land degradation.
- Farmer-to-farmer exchange visits are very effective in stimulating the horizontal spread of (SLM) community initiatives, as demonstrated by the high adoption rates. This may be associated with the practical field level interactions in technologies and practices, and the free and open flow of information among farmers.

- Sensitisation about and mainstreaming of SLM should be based on visual demonstration of practices by land users who can testify the results and benefits.
- SLM needs an enabling environment – investments are especially needed in those infrastructures that contribute to improved rural livelihoods (e.g. roads, schools, hospitals) and better market access.

The SCI-SLM project was instrumental in uncovering the three community initiatives presented in this chapter. SCI-SLM identified, characterised and validated them, but most of all it contributed to the empowerment of the communities and the adoption/adaptation of their innovations. The project aimed at encouraging the communities, acknowledging the value of their initiatives, building partnerships and promoting the sharing of experiences. The SCI-SLM approach also enhanced the interaction between communities, researchers, practitioners and policymakers in a mutual learning undertaking. As a result of the positive experience, the approach is now being widely applied in the implementation of community-led activities by other SLM projects in Uganda – a concrete sign of institutionalisation of the SCI-SLM approach.

Notes

1   Uganda has maintained a high population growth rate at about 3.2 per cent per annum (NPA 2010).
2   Over 85 per cent of Uganda's population (about 27 million people) is rural and derives its livelihood from natural resources (UNFPA 2009).
3   One of the four types of land tenure systems recognised by the Constitution of Uganda, based on which tenants pay a fee to a landlord for the use of land and to have limited rights over it.
4   PROLINNOVA = 'Promoting local innovation' (www.prolinnova.net).
5   Night *kraaling* is a technique used to rehabilitate land by confining livestock overnight in an enclosed open-air space in a degraded area for a period of 2–3 weeks in order to enable continuous deposition of animal dung. Grasses establish. A new site for a *kraal* on denuded land is then chosen. The accumulation of rich organic matter in the soil improves its structure, nutrient content and moisture retention capacity, thus supporting the rejuvenation and vigorous growth of vegetation.
6   The in-field SLM practices, including conservation agriculture, used by the BANDERA community were also inspired by the experiences and lessons learnt from the Promoting Farmer Innovation project (see Chapter 2).
7   Permanent planting basins are constructed mainly for annual crops (cereals and legumes).

References

Adebowale, K. and Adedire, C. (2006), Chemical composition and insecticidal properties of the underutilized jatropha curcas seed. *African Journal of Biotechnology* 5, pp. 901–906.
Critchley, W., Miiro, D., Ellis-Jones, J., Briggs, S. and Tumuhairwe, J. (1999a), *Traditions and Innovations in Land Husbandry: Building on local knowledge in Kabale, Uganda*. RELMA, Nairobi, Kenya.

Critchley, W., with Cooke, R., Jallow, T., Lafleur, S., Laman, M., Njoroge, J., Nyagah, V. and Saint-Firmain, E. (1999b), *Promoting Farmer Innovation. Harnessing local environmental knowledge in East Africa*. RELMA and UNDP, Nairobi, Kenya.

Drechsel, P., Gyiele L., Kunze, D. and Cofie, F. (2001), Population density, soil nutrient depletion, and economic growth in Sub-Saharan Africa. *Ecological Economics* 38(2), pp. 251–8.

McFarlane, J.E. (1984), Repellent effect of volatile fatty acids of frass on larvae of German cockroach, Blatella germanica (L) (Dictyoptera: Blattelidae). *Journal of Chemical Ecology* 10, pp. 1617–22.

Ministry of Agriculture, Animal Industry and Fisheries (MAAIF), (2010a), *Uganda Strategic Investment Framework for Sustainable Land Management 2010–2020*. Kampala, Uganda.

Ministry of Agriculture, Animal Industry and Fisheries (MAAIF), (2010b), *Agriculture Sector Development Strategy and Investment Plan: 2010/11–2014/15*. Entebbe, Uganda.

Mutunga, K. and Critchley, W. (2001), *Farmers' Initiatives in Land Husbandry: Promising technologies for the drier areas of East Africa*. Regional Land Management Unit (RELMA), Nairobi, Kenya.

National Environment Management Authority (NEMA), (1995), *National Environment Action Plan*. Kampala, Uganda.

National Environment Management Authority (NEMA), (2005), *The State of Environment Report for Uganda 2004/05*. Kampala, Uganda.

National Planning Authority (NPA), (2010), *National Development Plan 2010/11–2014/15*. Ministry of Finance, Planning and Economic Development, Kampala, Uganda.

United Nations Population Fund (UNFPA), (2009), *The State of World Population 2009*. New York.

Reij C. and Waters-Bayer A. (2001), *Farmer Innovation in Africa: A source for inspiration for agricultural development*. Earthscan, London.

Voortman, R., Sonneveld, B. and Keyzer, M. (2000), *African Land Ecology: Opportunities and constraints for agricultural development*. Working Paper 37. Centre for International Development (CID), Harvard University.

World Bank, (2008), *Uganda – Sustainable land management public expenditure review (SLM PER)*. World Bank, Washington, DC.

9 Cross-learning with community initiatives

Wendelien Tuijp, Saa Dittoh, Mohamed Mahdi and Maxwell Mudhara

Introduction

Cross-learning is a process whereby individuals of groups exchange knowledge and information, and share ideas with each other, resulting in increased knowledge amongst those involved. This chapter discusses the process of cross-learning that played a central role in Stimulating Community Initiatives in Sustainable Land Management. SCI-SLM focused on identifying innovative forms of land management within communities in four African countries. The project helped to add value to their practices and encouraged the communities to learn from each other, principally by exchange visits – a particular type of cross-learning, where communities visit each other to exchange knowledge and ideas. The concept is founded on the premise that community initiatives will be strengthened by enabling community members to learn from each other. Usually farmers in rural African communities do not see themselves as experts; they expect outsiders to be 'experts': extension workers and researchers who tell them how to improve their SLM practices. This has been the predominant paradigm in development. On the contrary, SCI-SLM promoted the exchange of farmers' own ideas, experiences, knowledge and know-how.

It could be asked whether exchange visits can have a real impact? Can these short visits be more than just courtesy calls? Are the different geographical, cultural and agricultural circumstances not so diverse that they hamper the exchange of knowledge and information? SCI-SLM was based on the idea that visiting each other and seeing what each other are doing – and thinking – is such a strong experience that it is able to outweigh any differences. Farmer-to-farmer exchange visits have been successfully implemented in several projects on farmer innovation in Africa. For example the projects, Réseau Marp in Burkina Faso and Sahel Eco in Mali, have facilitated exchange visits between farmer groups within, and across, countries (reseaumarpbf.org; saheleco.net). They work with innovators on farmer-managed natural regeneration. Réseau Marp has also facilitated visits of northern Ghanaian farmer groups to cross-learn from community initiatives in Burkina Faso.

Under SCI-SLM the country teams played a key role in planning and facilitating the exchange visits, which were organised both *within* and *between* countries (see Chapter 3 also). The objective of the exchange visits was to enable community members to improve the quality and effectiveness of their initiatives by learning from each other. A further premise was that communities *without* SLM initiatives would become inspired and take up initiatives that were suitable to their own situations and contexts. They could learn the actual technology or social arrangement – and/or become more creative themselves. Exchange visits were intended to benefit both visiting and hosting participants through an open exchange of ideas, knowledge and sound practices.

The exchange visits within SCI-SLM created unique opportunities for sharing knowledge and cross-learning (Di Prima *et al.* 2013). Cross-learning through exchange visits was built into SCI-SLM from the start and it played a central role throughout the project. In general, community members selected the members who would join the visiting team during the cross-visits; the hosting community consisted of most members (see details in the overview Tables 9.1–9.4). In Ghana the country team organised within-country visits, consulting both communities to plan a date, and then organised transport to convey visiting members to the host community. After the usual greeting protocols, including paying courtesy calls on chiefs, fields were visited, initiatives shown and exchange of ideas and knowledge flowed. Reporting back occurred in several ways: formally during regular community meetings, or in meetings arranged specifically for this purpose. It happened in more informal ways also. In Morocco the reporting back was often more spontaneous than systematically organised.

Exchange levels and themes

Organising and facilitating exchange visits was one of the key activities for the SCI-SLM country teams: this helped to create powerful – though informal – learning platforms for participating communities and other stakeholders. The project asked researchers to take a different approach from business-as-usual by becoming attuned to the needs and wishes of the community members and to join hands in 'joint experimentation' (Critchley 2007). After the selection of four community initiatives in each country (see Chapters 5–8), exchange visits were organised throughout the project period. Cross-visits were either bilateral, when the exchange occurred between two communities, as was the case in Ghana and Uganda or multilateral, as in the case of most visits in Morocco, where members of several communities visited a community initiative together (see Chapter 6). As already touched upon, cross-visits were organised at different levels:

1   *Intra-country*: exchange visits between innovative communities within a country.
2   *Between country*: bilateral visits between innovative communities of two countries.
3   *Regional visits*: involving members of innovative communities from all countries, visiting SCI-SLM communities in the country hosting the Regional Steering Committee (RSC) meetings.
4   *Study visits*: communities visiting research stations and other relevant project sites to learn specific SLM techniques.

As explained in Chapter 1, SCI-SLM brought together four countries with different settings and conditions, including the backgrounds of the lead partners and their experiences on farmer innovation. Because of these differences, the cross-learning process had different starting points in each country. Both Uganda and Ghana already had experience in working with indigenous knowledge and farmer innovation. The BANDERA community in Uganda was inspired by the Promoting Farmer Innovation methodology (Chapter 3), which had stimulated awareness of their own capacity as innovators. In addition, SCI-SLM Uganda was coordinated by the Ministry of Agriculture, Animal Industry and Fisheries, and in this country the project was already up and running on national counterpart funding before the official start in 2009. The University for Development Studies in Ghana has a mandate to integrate indigenous knowledge with scientific knowledge at community levels, and its students are required to live in rural communities as part of the curriculum, in order to understand indigenous knowledge and community life. Appreciation of indigenous knowledge by community members themselves has, however, been weak in most parts of Ghana. In South Africa there were some experiences with farmer innovation and indigenous knowledge through the PROLINNOVA programme (Prolinnova.net). In Morocco, the coordinating agency, Targa-Aide, had no specific experience with farmer or community innovation. Overall, the unique multi-country set-up of SCI-SLM created valuable opportunities for all participating countries to promote cross-learning. It allowed flexibility in the design and organisation of the cross-visits, and each country team refined its own strategy.

A. Intra-country (inter-community) exchange visits

During intra-country exchange visits, members of an innovating community visited another innovating community within the same national borders. In some cases a community without an initiative visited an innovating community. This last type was categorised under study visits (see above). An important objective of the community exchange was to improve the existing initiatives on SLM and strengthen the capacity of community members. During SCI-SLM a total of 11 intra-country exchange visits were organised, four in Uganda, three in Morocco and two each in Ghana and South Africa (Table 9.1).

During the cross-visits SCI-SLM helped to create an open, friendly atmosphere in which all participants felt free to express their expectations and to actively take part in group discussions, field visits and sampling. In all four countries this was facilitated by the welcoming nature of African communities. In Morocco this was because of the long-term relationships based on mutual respect, confidence and exchange of hospitality, which is part of the culture and a non-written rule. Hospitality in Ghanaian tradition is the rule too; people feel honored to receive visitors who have put considerable effort into their travel and visit. Such a spirit led to friendship and companionship between community members and it engendered positive effects on the confidence of the community members.

The exchange visits also gave the often isolated community members an opportunity to see and experience the 'outside world' and to get to know fellow men and women in other parts of their country. In some cases it opened a connection with more enclosed areas, for example the communities of Ouneine in the High Atlas of Morocco. In other cases communities were relatively close to each other - without knowing of each other's practices. In Ghana the communities of Kandiga and Moatani both used composting techniques to increase soil fertility but were not aware of each other even though they were quite close.

Ghana

The University for Development Studies aims to integrate indigenous knowledge and cultural practices of local communities with scientific knowledge (see Chapter 11). Its 'plug-in' principle has been developed to engage with and support rural communities to improve what they are already doing themselves. This fitted seamlessly with the SCI-SLM approach, and UDS organised two intra-country (inter-community) cross-visits between Kandiga-Atosali and Moatani communities. Both communities practise composting to improve soil fertility, but using different techniques (see Chapter 5 for details). Moatani is approximately 80 kilometers from Kandiga-Atosali, yet none of the members in either communities had any knowledge of the technology of the other – and they were independently addressing common problems of infertile soils. They only got to know of the different technologies during the exchange visits - and that widened their innovative horizons. SCI-SLM Ghana worked towards making them complement each other.

The cross-visits created extensive and constructive peer-to-peer learning experiences for all participants, including scientific researchers who examined the composting innovation of the Kandiga-Atosali community. Both communities inspired each other, with the Moatani composting pits being more labor intensive than the heaps in Kandiga-Atosali, but richer in materials composted and also faster and better decomposing due to the continued watering. The women of Moatani learned the technique of compost heaps by their peers from Kandiga-Atosali, and started to experiment in their own fields. The Kandiga-Atosali community members were impressed by the social organisation of the Moatani initiative. The women's groups of Moatani explained the importance of common interest farmer groups: they help and learn from each other, which results in more unity between the women than before. The two cross-visits established new relationships between the communities. It was observed by the country team that, during the return visit, the interactions were more friendly and characterised by greater familiarity. Possibly individual members made friends across the communities and discussed a range of issues during the return visit; but there is no documented evidence of exchanges between the different community members outside the project visits.

Morocco

For decades Targa-Aide had been working with communities, but the concept of farmer innovation was new to the organisation. Three of the selected initiatives were located in Ouneine, an isolated valley in the High Atlas mountains, where people had very limited contact with their peers outside the valley (see Chapter 6). These conditions had an impact on the organisation of the cross-visits. The initial intra-country exchange visits offered opportunities to bring the communities together, present the selected initiatives and the SCI-SLM project with its methodology to community members, and to build common understanding of its goals.

Meetings and workshops with community representatives were organised with presentations, roundtable discussions, field visits and reflections on lessons learnt. Several different environments and community initiatives were covered, and joint evaluation of progress on the development of initiatives was established. In total Targa-Aide organised four exchange visits between the Moroccan communities (see Table 9.1 for details). In most cases these were multilateral exchange visits in which representatives of all selected initiatives participated. The multilateral base made these visits unique and rich peer-to-peer learning experiences.

Peer-to-peer advice and feedback were exchanged on various topics, such as the treatment of tree diseases, the cultivation of beans, growing fruit trees, e-commerce, different modes of forest management, best practices and common challenges. The diversity of activities practised by the Ighrem Association in Tabant - located in a less isolated part of the Atlas where there is tourism - was a clear source of inspiration and an example to be followed by the communities in Ouneine. Community members explained their initiatives using photos, power point, videos and sketches to illustrate their stories and to add value to the presentations of their initiatives.

South Africa

The South African SCI-SLM team organised three cross-visits in which forest management was the central theme. Community members from New Reserve B and KwaSobabili visited Mathimatolo community. Farmers exchanged knowledge on optimal wattle (*Acacia meurnsii*) densities to produce sufficient tree products, meanwhile safeguarding its sustainability for future use. It included thinning, pruning and staggering tree planting, seed collection and treatment (Chapter 7). Visiting community members realised that pruning and replanting wattle trees is important for the survival of the forest. The KwaSobabili community board decided to allow community members to prune trees and harvest wood from older trees, and to collect seeds. As a result the community was able to successfully manage their forest resources and they transformed the uncontrolled growth into a managed forest, while protecting their water resources. It created employment opportunities, which increased the motivation of community members to actively protect the wattle forest. The KwaSobabili initiative flourished and started to attract attention from people in surrounding communities.

In New Reserve B there were conflicts amongst community members about the need to maintain the forest; some people sold the wood illegally to neighbouring communities, which resulted in its decline. Researchers of the University of KwaZulu-Natal created an opportunity for them to visit the KwaSobabili initiative. After the cross-visit, the New Reserve B community decided to install a working group spearheading sustainable forest management on behalf of the community; this was an important social achievement. A key aspect of the cross-visit was the informal interaction that allowed all participating community members to mix with each other and exchange on a one-to-one basis.

Uganda

In Uganda the conditions to promote community initiatives were relatively favorable. SCI-SLM was coordinated by the Ministry of Agriculture, Animal Industry and Fisheries (MAAIF), that was already experienced with farmer innovation through the precursor of the SCI-SLM project – Promoting Farmer Innovation (PFI) (see Chapter 2). MAAIF also had the responsibility to implement SLM policies nationally, and decided to begin activities before the official SCI-SLM start up in 2009. In total the country team organised four bilateral exchange visits between the participating communities.

Members of BANDERA and NACIA exchanged ideas and knowledge on night *kraaling* techniques for cattle, conservation agriculture and water harvesting technologies. Farmers of NACIA had established citrus orchards, and this was of great interest to the BANDERA participants. A second technique that caught the eye of BANDERA members was the use of underground water tanks for drip irrigation, household use and watering livestock. In return the BANDERA members explained conservation agriculture (CA) to their peers, with its constituent components of minimum tillage, mulching, crop rotation and agroforestry.

The RECPA and Banyakabungo communities met each other twice. They compared their ideas and shared knowledge on the integration of livestock into farming systems, forest management and tree planting. These farmer-to-farmer visits provided important occasions to exchange relevant technologies and know-how in a simple, practical and cost-effective way, with the project team stepping aside and allowing the community members to take center stage.

B. Between country exchange visits: bilateral

SCI-SLM being set up in four different African countries created opportunities to organise cross-border exchange visits as well. The exchange of knowledge between countries was an important part of the cross-learning and strengthened the implementation of the project. International exchange visits were either between two countries or involved all SCI-SLM countries during Regional Steering Committee meetings.

International exchange visits are more demanding in terms of organisation and obviously required a much higher budget provision. Considering the modest overall project funding, it is not surprising that only one bilateral visit took place in June 2012: a team of four Moroccan community representatives, together with two team members from Targa-Aide and one staff member from Vrije Universiteit Amsterdam, visited three community initiatives in northern Ghana. The trip was complemented by a study visit to a tree nursery.

The main goal of this visit was sharing and exchanging ideas and experiences of Moroccan community representatives with their peers in northern Ghana (see Figure 9.1). This cross-visit turned out to be inspirational and exciting. The Moroccan community members were very interested in the initiatives of Kandiga-Atosali, Moatani and Zorborgu communities. During three days of field visits there was enough time to have fruitful exchanges of experiences and sharing of ideas on

Table 9.1 Overview of SCI-SLM intra-country cross-visits

Country	Location/hosting community	Visitors	Shared topics	Community innovation adopted/lessons learned	Date visits
Ghana	Kandiga–Atosali (12 women, 8 men)	Moatani women's group (15 women)	Different composting techniques	Less tedious composting technique	Sep 2011
	Moatani (18 women, 10 men)	Kandiga (10 women, 6 men)	Different composting techniques	More efficient composting technique; strong social structures	April 2014
Morocco	Afourigh, Anzi, Lamhalt (61 women, 39 men)	Agouti (3 men), TAG team	Exchange and sharing of all present initiatives	Idea to set up a website for commercial purpose; treatment of tree diseases	May 2011
	Agouti (11 men)	Anzi, Afourigh, Machal, Tigouliane (10 men)	Exchange and sharing of all present initiatives	Diversification of activities; environmental awareness	Sep 2011
	Afourigh, Anzi, Lamhalt (61 women, 39 men)	Agouti, Tigouliane (6 men)	Exchange and sharing of all present initiatives	Importance of good relations and collective commitment, new horizons	Dec 2012
South Africa	Mathimatolo community, Greytown	KwaSobalili, Reserve B (9 women, 12 men)	Wattle tree management	New tree management techniques	June 2011
	Amazimeleni community	Gudwini community (11 women, 3 men)	Management of indigenous forest trees	Seed collection and tree nurseries	Feb 2012
Uganda	BANDERA community (23 women, 22 men)	Nalukonge community (4 women, 9 men)	Conservation agriculture; In-field SLM practices	CA, combined with soil and water conservation structures, improves resilience and production	Sep 2011
	Nalukonge Community (NACIA) (20 women, 9 men)	BANDERA community (4 women, 8 men)	Land rehabilitation; kraaling techniques, water harvesting tanks	Livestock/crop integration improves soil fertility; water harvesting tanks	Oct 2011

Table 9.1 Continued.

Country	Location/hosting community	Visitors	Shared topics	Community innovation adopted/lessons learned	Date visits
Uganda	RECPA community (8 women, 12 men)	Banyakabungo community (3 women, 10 men)	Tree management and tree nurseries	Tree nursery establishment; biogas; apiary; poultry	Oct 2011
	Banyakabungo (15 women, 10 men)	RECPA community (4 women, 9 men)	Livestock integration; agroforestry	Agroforestry, increasing social, economic, environmental benefits	Oct 2011

Source: Ghana, Morocco, South Africa and Uganda SCI-SLM teams.

social and technical issues between the Moroccan visitors and their Ghanaian peers; it is not known whether these ideas have materialised – once again pointing out the importance of post-project follow-up to assess impact.

The Moroccan visitors were rightfully impressed by the social organisation of both the women's groups of the Moatani and Zorborgu communities. In Zorborgu, people had initiated several activities to combat land degradation, such as protecting and planting trees, and non-burning of residual biomass in the dry season. As a result their crop yields and livestock numbers have increased, and their shea nut trees yield more nuts. The visitors from Morocco were impressed by the number of related, diverse activities to improve both the environment and people's lives. Key to the success of their integrated approach was the strong social structure, with different committees responsible for specific SLM topics, and the respect towards the Chief and his role as supervisor. The social organisation of Zorborgu is an innovative blend of traditional and modern decision-making processes. The visitors were touched by the sense of solidarity, commitment and mutual respect.

The initiative by the women of Moatani was also characterised by strong social organisation. They bundled their energies and ideas to solve several related challenges, such as intensive labour, to dig their compost pits. The two women's groups established a micro-finance system as well, to help each other and create more financial flexibility. Overall it had a positive impact on their yields and their contribution to household income. Diversifying and increasing income is high on the agenda of most community members, and the women of Moatani try to do this through the production and marketing of shea butter. In Morocco women produce Argan oil, which is of parallel value to the women. But, overall, the women of Moatani were more involved in commercial activities than their Moroccan counterparts. The Moatani women's groups and their strong social organisation were a source of inspiration for many visitors, especially because they were women without formal education. It helped to raise the awareness of the Moroccan male farmers about the role of women in community activities, and it also possibly stimulates the involvement of more women; but there is no hard data available.

The different composting techniques that Kandiga-Atosali and Moatani communities used to prevent soil fertility depletion were significant to the Moroccan visitors. The Moroccan community members appreciated this organic way of enriching the agricultural soils as an alternative to chemical fertilizers, the latter being practised in their own communities. This was one of the ideas taken back to Morocco, but it is not yet known whether people actually started to practise this initiative. The idea of community tree nurseries convinced the members from Tabant community to restart their own project in cooperation with the High Commission for Water and Forests, which was still waiting to be implemented. Water scarcity is an issue in Kandiga-Atosali, and in this context the farmers from Morocco explained the possibility of rooftop harvesting and designed a system on paper.

Even though the visitors from Morocco and those in Ghana live under very different geographical, cultural and agricultural circumstances, they share one crucial common denominator: they are land users and innovators, and in that perspective they speak the same language. This paves the way towards cross-learning (Critchley 2012).

Figure 9.1 Moroccan delegates meet with farmers from Kandiga, Ghana – June 2012

Figure 9.2 Memunatu Seidu from Moatani, Ghana explains the composting technique during the international workshop, Ghana – September 2012

Table 9.2 Details of international exchange between Ghana and Morocco

Hosting communities (Ghana)	Visiting communities (Morocco)	Topics	Lessons learned	Date
Moatani, Kandiga, Zorborgu (about 34 women, 24 men)	Afourigh, Agouti, Anzi, Tigouliane (4 men)	Composting, non-burning, tree planting	Ideas for composting, tree planting, women as agents of change	June 2012

Source: Ghana and Morocco SCI-SLM teams.

C. Regional visits

SCI-SLM's regional approach, with community initiatives practised in the four corners of Africa, offered a unique opportunity for cross-learning experiences. In addition to the bilateral exchange visits there were the Regional Steering Committee meetings, in which participants from all four countries joined to exchange experiences, share knowledge and update each other on on-going activities. The RSC meetings were an obvious opportunity for international exchange in that partners and participants from all countries met once a year. From the start, Targa-Aide in Morocco was particularly interested in the regional approach of SCI-SLM and the aspect of South-to-South learning. Projects that include both North African and Sub-Saharan African countries are not very common, and SCI-SLM provided the opportunity to bridge that gap.

Each country hosted a Regional Steering Committee meeting. During two of these international cross-visits, community representatives from visiting countries participated as well. The other participants were the national team members from the different countries, TAG representatives, the portfolio manager from UNEP/GEF and several stakeholders from the hosting country, such as members from the national steering committees, representatives from (local) government, and others. Highlights during these annual meetings were the field visits to the local community initiatives. The communities warmly welcomed all visitors, creating an informal atmosphere. Despite the differences in language, culture, geography, gender and perceptions, participants interacted very effectively and constructively (as illustrated in Figure 9.2). It enabled the establishment of new friendships and relationships. The hosting communities became aware that they were doing interesting things that attracted many different stakeholders. This awareness was very encouraging, and the effect of the annual Steering Committee meetings with many international visitors was even stronger.

Beyond the exchange of technical and social knowledge, the international visits were an opportunity to learn from each other about culture, including cooking, singing, dancing and lifestyle, broadening horizons to diversity and stimulating respect for the culture of others. This was an irreplaceable aspect of exchange visits.

Study visits

Overall it is clear that all cross-visits included study components to some degree, a reason why the lead partner in Morocco considered all exchange visits to be study

Table 9.3 Overview of regional visits

Hosting country	Location, hosting communities	Visitors	Shared topics	Lessons learned	Date
South Africa	CEAD, Pietermaritzburg	SCI-SLM country teams, TAG team, UNEP	Exchange and share ideas on project and methodology	Increased understanding of principles and ideas of the project	Nov 2009
Morocco	Afourigh, Anzi, Lamhalt (61 women, 39 men)	No community members from other countries present	Land rehabilitation; water use rights; wood craft	Increased awareness of hosting communities on the importance of their innovations	Sep 2010
Uganda	Bandera, Recpa, Banyakabungo (73 women, 47 men)	Four community members from all other countries	CA; reforestation and tree management; livestock integration	There is need to properly document and share community initiatives	Oct 2011
Ghana	Moatani, Kandiga, Zorborgu	Community members from all other countries (1 woman, 5 men)	Composting; non-burning; tree planting	Ideas for composting; dedication of women as agents of change	Sep 2012

Source: SCI-SLM Coordinating Office, UKZN.

tours as well. But this section only describes the study tours aimed at specific study purposes, in most cases technical aspects. SCI-SLM facilitated these study visits throughout the project period. Community members visited research stations or other relevant project sites to learn how to practise specific SLM techniques. In some cases researchers, including the community member experts, visited the communities to demonstrate and explain new SLM techniques. Sometimes study visits led to joint experimentation; in Uganda two joint experiments have been developed by farmers in cooperation with researchers (see Chapter 8).

Most study visits and training opportunities were organised in Uganda (Table 9.3). In 2004 NACIA community members learned about water harvesting using an underground tank and biogas use through a study tour to south-western Uganda. When BANDERA community members visited the NACIA community in 2011, farmers from NACIA had become experts themselves. As a follow-up to this exchange visit, a member of NACIA trained BANDERA farmers in the construction of underground water harvesting tanks. In Kamwenge district, NACIA community members gained new knowledge and practices about termite control. The BANDERA community members learned about fruit tree management and agroforestry, and specifically about conservation agriculture in Iganga, Pallisa and Soroti. They became experts themselves and have subsequently trained many other farmers on CA. The farmers of the RECPA community negotiated a partnership with NAADS to be trained on high-yielding crop varieties. Researchers from Makerere University trained RECPA members on the use of a simple field soil-testing kit for field soil analysis (see Chapter 8).

In the other countries, the organisation of study visits was on a lower scale. In Ghana the farmers of Kandiga-Atosali were happy and proud that researchers showed interest in their innovation on composting. Farmers from other rural areas in Ghana were taken on study tours to Kandiga-Atosali to learn about this technique.

In South Africa a delegation of several departments of the Uganda Government visited Gudwini community to learn more about indigenous forest rehabilitation. This fitted with the purpose of their international study tour: learning more about tools for sustainable land management. The KwaSobabili community learned, in Greytown, more about technical aspects of wattle forest management. When community members of KwaSobabili had become experts themselves, SCI-SLM organised a field day for neighbouring communities who were interested in wattle forest management. Community members from Reserve B who had a similar initiative also participated in the field day.

Stimulating community initiatives

In all four countries, SCI-SLM stimulated community initiatives through exchange visits, specific study visits and linking the communities to other partners and networks, including local government on regulations and legal aspects. SCI-SLM promoted the use of indigenous knowledge in providing sustainable solutions to land degradation by supporting partnerships between communities and research institutions to bridge the knowledge gap, thus generating 'hybrid knowledge'.

A significant role of SCI-SLM was supporting communities to realise the importance and the value of their initiatives, since this was not evident for most community members. Visits to the various sites by scientists and farmers from within and outside the country gave community members added confidence in their initiatives. There has, without doubt, been considerable benefit to community members as well as researchers and extension agents in terms of greater belief in community ability to be innovative and solve problem themselves.

Most study visits have taken place in Uganda, which led to joint experimentation in a number of cases. Two joint experiments were set up, as described in the Uganda chapter. Other countries organised some study visits as well, but joint experimentation was generally weak.

Lessons learned by different stakeholders

During cross-visits, farmers, researchers, extension agents and other visitors interacted informally but very effectively. Even within a single day, community members

Table 9.4 Overview of study visits

Country	Location – hosts	Visitors	Topics	Lessons learned	Date
Morocco	Tigouliane (355 women and men)	Agouti, Afourigh, Anzi (4 men), the trainers	Fruit tree planting; environmental awareness	Tree planting, land rehabilitation	Feb 2012
South Africa	KwaSobabili (78 women, 55 men)	Reserve B, Mathamo (34 women, 41 men)	Wattle forest management	New management techniques and social organisation	Nov 2012
	Potshini, Bergville	Amavimbela (2 women, 33 men)	Holistic grazing management; no-till planting	No-tillage techniques; social organisation	March 2013
	SilverGlen Nature Reserve (2 women, 3 men)	Gudwini Msinga (16 women, 12 men)	Management of indigenous forest trees	Techniques on seeding and seedling management	April 2013
Uganda	Iganga, Pallisa and Soroti (12 women, 8 men)	Bandera community (3 women, 10 men)	Fruit tree management and agroforestry	Expansion of area under fruit trees; water-harvesting basins; mulching; livestock integration	April 2011
	Bandera community	NACIA trainer	Construction of underground water-harvesting tanks	Ability to construct water-harvesting tanks	Nov 2011
	Pallisa	BANDERA community	Study on conservation agriculture	Improved knowledge on CA; capacity to train others	Dec 2011
	NARO, Kamwenge district	NACIA community	Use of arboreal termites to control terrestrial termites	Knowledge about the use of arboreal termites	Oct 2012

Table 9.4 Continued.

Country	Location – hosts	Visitors	Topics	Lessons learned	Date
	RECPA community	NAADS trainers	Use of soil-testing kit	Efficiently manage the arable soils	June 2013
Intnl	South Africa, Gudwini (37 women, 26 men)	Several government departments of Uganda incl. SCI-SLM team (7 women, 20 men)	Indigenous forests	Learning about SLM practices in South Africa	July 2013

Source: Morocco, South Africa and Uganda SCI-SLM teams.

learned so much from each other that they were able to experiment on the basis of each other's initiatives. According to the Ghana country team, experience shows that exchange visits between farmers have a greater impact than a series of workshops and visits by extension agents. In fact the knowledge gained through peer learning could never have been transferred by any extension worker or researcher. It is entirely feasible that community members will come up with new initiatives as a result of their improved and widened knowledge. In Ghana there was evidence in both Kandiga-Atosali and Moatani communities of people trying the initiative of the other community. In all likelihood, over time they will come to understand the advantages and disadvantages of each composting technique and either integrate them, or develop other initiatives based on the experiences they have gained from trying the two initiatives. The exchange visits also established close relationships between members of different communities, as observed by the Ghana country team.

Box 9.1 Specific lessons learned and initiatives adopted by communities in the four countries

Ghana

- Kandiga-Atosali community members learned from the Moatani women that strong social organisation is crucial in promoting farming activities. Currently there is a thriving SLM group in Kandiga-Atosali.
- The Moatani women learned about anther composting technique, compost heaps, by visiting the Kandiga-Atosali community. After their return home several women started to experiment with this type of composting.

Morocco

- Cross-visits to Äit Bouguemez, Uganda and Ghana developed environmental awareness of the Anzi community members: they realise the importance of sustainable forest management. As a result, they linked up with the Forest department.
- The Tigouliane community (without an initiative) learned to appreciate the techniques of land rehabilitation as a result of a study visit to Ouneine and Tabant. Recently they started to rehabilitate their own land, with a subsidy from the Moroccan government.

South Africa

- A visit to Uganda taught Gudwini community representatives that you can take sustainable forest management to a higher level. This motivated the community to raise tree nurseries with collected seeds of endangered indigenous trees and then sell or plant seedlings.
- Kwasobabili community learned the importance of pruning and replanting of wattle trees to guarantee the forest for the next generations. The harvested wood and collected seeds created income and helped the community to successfully manage their forest resources.

Uganda

- BANDERA community members learned about the use of underground water harvesting tanks constructed in NACIA community. They adopted the construction with technical guidance of NACIA community members. Since then, BANDERA members have constructed 16 water harvesting tanks.
- The Banyakabungo grazing association picked up tips about stocking rates/rotational grazing and invasive alien species (in this case *Lantana camara* and *Opuntia spp*) from Moroccan community members.

Outscaling – other communities taking up initiatives

One of the targets of SCI-SLM was the spread of community initiatives to other communities. This horizontal spread of SCI-SLM initiatives demonstrates one impact of the farmer-to-farmer exchange visits. Generally it was easier for country teams to capture qualitative rather than quantitative data; estimates of the number of hectares proved to be especially difficult to record.

Starting points in all countries were different, and this had an impact on the results of the spread of the initiatives too. It appears that the uptake of initiatives by new communities is highest in Uganda, with an enabling environment, a history of strong commitment and social networking in the case of the BANDERA community. BANDERA members have become experts on conservation agriculture using several strategies to spread CA, including setting up demonstration plots and a Model Farmer Resource Centre, where they display various land management practices. The community established an outreach training team as well. By the end of 2013 the team had reached out to 315 farmers, carried out about 31 training sessions, and had set up over 100 conservation agriculture gardens.

Some initiatives were taken up by other communities spontaneously. In Ouneine, Morocco, for example, 18 families have adopted the initiative on land rehabilitation. About 18 plots of degraded land have been transformed into irrigated production systems. The rehabilitated land covers about 64.5 ha and the area is increasing each year. The composting techniques by Kandiga-Atosali and Moatani communities in Ghana proved to be very inspiring. It was not only communities in Ghana that started to practise this SLM technique after visiting Kandiga-Atosali and Moatani, but also community members from Morocco and South Africa also brought this idea back to their communities.

Country teams also actively stimulated communities without initiatives by organising exchange or study visits to innovative communities. The SCI-SLM team in Morocco organised an exposure visit for Tigouliane community to SCI-SLM communities. Tigouliane community showed interest in land rehabilitation. Tigouliane members also started to plant fruit trees after visiting their peers in Agouti community.

In some cases the uptake evolved in a more complex way. In South Africa the KwaSobabili initiative thrived and was noticed by surrounding communities, who then approached KwaSobabili members to learn more about their initiative. This convinced SCI-SLM to facilitate a field day where other communities, local government and traditional council were able to learn more about the initiative. Two neighbouring communities of Reserve B, Reserve A and Reserve C, also started proactively managing their forests. SCI-SLM linked these two new communities to Rural Forest Management, a private company that assists rural communities in setting up business plans for production of wattle. It is hoped that this linkage will translate into employment creation in both communities.

Not all flourishing and inspiring initiatives were taken up by other communities, as was proven by the initiative of Afourigh community on water use rights in Morocco. It was evident that the initiative had resulted in increased crop production and crop diversity. Although many visitors were interested by this social innovation on water management, initiated by a group of young men, the initiative did not spread to other communities. Water use rights are considered sacred and untouchable, which explains one difficulty in extending this initiative to other communities.

The facilitation of farmer-to-farmer visits within, and also between, countries is also good preparation for future up-scaling programmes. This was demonstrated by the cross-visit between Morocco and Ghana. According to both country teams, it is a powerful way to encourage exchange and adoption of progressive thinking and improved practices. The idea of stimulating existing innovations to achieve results on larger scales is an achievement, even though considerable effort is still needed in most of the participating countries (Kojwang 2013).

Improved knowledge and capacity building

According to Kojwang (2013) in the Mid-term Review, 'the potential of local innovations can be fully exploited only when: i. innovators become aware of the value/importance of their innovation; ii. there is an enabling/conducive environment; iii. local and scientific knowledge merge (hybrid knowledge)' (Kojwang 2013). The exchange visits addressed two of these conditions. At the start of SCI-SLM the awareness of the value of the innovations was not evident in many communities. The exchange visits helped to build confidence and enhanced appreciation by community members themselves of the worth of their own innovation. This created a sense of reassurance, provided encouragement and strengthened their commitment at the same time. The hosting communities came to understand that what they were doing was interesting enough to attract different stakeholders, even from abroad. This awareness was in itself very stimulating to the hosting communities. The effect was stronger when international teams visited during the annual Steering Committee meetings. Community initiatives were improved by the community members after exchange visits and new practices evolved. In fact all exchange visits motivated communities to excel. Clearly these exchanges were beneficial both to the visiting and to the hosting participants.

For example, the exchange visits and the interest by visitors from outside made the women of Moatani (in Ghana: Chapter 5) realise that they were doing something really interesting, worthwhile and innovative. These women – without formal education – broke specific socio-cultural barriers too. The exchange visits helped to encourage the men of Moatani to take interest in the work of the women, and the men decided to assist the women more than they used to do. With growing self-confidence and enthusiasm, the women continued their activities, inspiring many others. Strong social structures and collective commitment are crucial to improve the livelihoods and combat land degradation. The women's groups of Moatani understood this very well and they inspired the Moroccan farmers during an exchange visit. As a result these farmers motivated more women in their communities to participate in community activities and initiatives.

The farmers in Kandiga-Atosali were impressed by the social organisation of the Moatani women as well. Even though composting in Kandiga was very beneficial to the community , they had difficulties in discussing farm problems amongst themselves due to lack of social organisation. After cross-visits, Kandiga-Atosali members were convinced about the importance of proper social organisation in the promotion of farming activities. Currently there is a thriving SLM farmer group in Kandiga-Atosali.

In BANDERA community, Uganda, farmer-to-farmer learning activities significantly contributed to their understanding of SLM, their organisation and self-motivation, as well as willingness to learn and train others. The community demonstrated strong social organisation, capable of mobilising a large small-scale farming community to collectively farm a larger area on a commercial scale. They embraced the government's export drive while simultaneously protecting the River Nile banks. This major achievement inspired many people including political leaders.

Reserve B community members in South Africa had a wattle forest management initiative, but experienced challenges as to how to organise themselves to manage the forest. During a field day in KwaSobabili, community members showed their peers from Reserve B how they worked, which allowed the people from Reserve B to improve the management of their own forest and to strengthen their initiative. The initiative in Reserve B inspired surrounding communities, which had a positive impact on the confidence and motivation of the people in Reserve B to continue their activities (Chapter 7).

Despite strong social organisation and dedication, Zorborgu community (in Ghana: Chapter 5) faced challenges in protecting the forest, mainly from hunters from neighbouring communities, and other difficulties, such as tree cutting and bush burning. But the cross-visits both by the Moroccan SCI-SLM team in June 2012 and by all country teams in September 2012 helped the community members resolve to continue their forest conservation activities.

Exchange of knowledge between community members, researchers and extension agents was very valuable, but in most cases they did not reach the stage of joint experimentation. The exception was Uganda, where communities and researchers were well ahead in merging local and scientific knowledge. The idea is that communities should cooperate with research stations and universities in joint experimentation to validate and improve the use of indigenous knowledge and its application to address land degradation and productivity challenges. SCI-SLM Uganda encouraged joint experimentation and provided, at least in theory, an effective platform for the interaction of the communities and research institutes (Chapter 8).

The inter-country visits have demonstrated that exchange of experiences between countries across the continent is a valid vehicle for addressing many similar agricultural and environmental problems at the local level. The effect of the annual Steering Committee meetings with many international visitors was especially strong. As stated in Chapter 5, it was very revealing that community members from the Atlas Mountains of Morocco could identify themselves with the problems of community members in northern Ghana during their one-week visit to northern Ghana and that an SLM practice in northern Ghana would be of great interest to community members in the Drakensburg area of South Africa.

During the regional exchange visit to Uganda, community representatives from South Africa realised the need to take their own initiatives to higher levels. Farmers in Uganda were not only growing fruit trees as a community initiative, but they continued to add value to the fruits they produced on their farms. A representative from Msinga was motivated not only to manage the natural forest sustainably, but to raise tree nurseries with collected seeds of the endangered indigenous trees, selling or planting the seedlings. This idea was adopted by the community (Chapter 8).

Exchange visits provided the opportunity to learn more about sustainability in several ways. Community members shared information on their technical and social innovation; in South Africa the KwaSobabili community learned how to manage their wattle forest more sustainably. Currently the community associates the forest with the creation of employment opportunities, saving about US$ 7,500 per year. The change in managing the forest by allowing people to harvest wood in a sustainable manner benefited the poor especially. The benefits also had a positive impact on the social cohesion of the community. The farmers in Ouneine learned from their peers in Uganda how to reverse land degradation and to make better use of forest resources in a sustainable way. They have become highly motivated to increase environmental consciousness in their communities.

Community members also compared (local) rules and regulations governing their initiatives, because local rules and regulations are crucial to the sustainability of their initiatives. It is important to be able to build partnerships between communities and authorities, and Uganda is an inspiring example in this respect. Furthermore improved knowledge through the exchange visits increased employment opportunities, which is particularly interesting for youth in the communities as it can prevent them going to look for work in the cities. KwaSobabili and Mathomo in South Africa are clear examples in this sense.

The project has in some cases brought out marginalised and remote communities to exchange with those from other countries on technical natural resource management and social issues, which hardly happens otherwise. For the community members in Ouneine in the High Atlas of Morocco, the exchange visits opened the door to their peers in the outside world with unprecedented opportunities to share and exchange experiences and knowledge on social and technical issues. It clearly widened their horizons and deepened their knowledge.

Working across countries that are far apart has challenges no doubt. Distant communication can be difficult and organising inter-country exchange visits demands a lot of effort. International exchange visits require intensive organisation and the costs are obviously high.

Conclusion

SCI-SLM's regional approach, with community initiatives practised in the four corners of Africa, has offered a unique opportunity for cross-learning experiences. The inter-country visits have demonstrated that exchange of experiences between countries across the continent is a valid vehicle for addressing many similar agricultural and environmental problems at the local level. Even though the geographical and cultural circumstances are different, all community members are land users and innovators in land management, and in that respect they speak the same language. Moreover, in all countries the exchange visits between the communities created an important platform for farmer-to-farmer learning experiences, exchange of valuable knowledge and sharing of ideas for all participants, including researchers. The exchange visits were rich experiences of great value: within one day farmers learned so much from each other and became inspired to the extent that they started to experiment with each other's initiatives (see Box 9.1). It is believed that community members will come up with new initiatives as a result of their improved and widened knowledge.

The exchange visits had a positive impact on the confidence of the innovators too. As pointed out by Kojwang in the Mid-Term Review, for a local innovation to be fully exploited it is necessary that innovators are aware of the value of their innovation (Kojwang 2013). This is exactly where the exchange visits in SCI-SLM have served the project's purpose well, because at the start of the project in Ghana almost no community members realised the importance of their initiative. They were surprised that educated people were interested in their simple and 'backward' technologies that – they considered – were merely to ensure their survival. Through the exchange visits, hosting community members came to understand that what they were doing was interesting and valuable, and this built their confidence and motivation to continue and excel. This effect was even stronger during the visits by international teams. The Ghana country team observed this general attitude in many communities with local initiatives. Unfortunately there are no statistics captured to quantify this.

Strong social structures and collective commitment are crucial to fight land degradation and improve livelihoods of people in rural African countries. The community members of BANDERA in Uganda and the women of Moatani in Ghana have shown this very well. They were great sources of inspiration to many peers whom they met during the cross-visits, who claimed they would start to pay more attention to the social structures of their communities. The exchange visits also established strong relationships between members of different communities, as observed by the country teams. Beyond the exchange of technical and social knowledge, the international cross-visits were an opportunity to learn from each other about culture.

Did the exchange visits result in the adoption of the newly learned practices by the communities? There is no simple yes or no answer to this question, due to the complex, dynamic and ongoing nature of learning processes. Learning comprises multiple steps: a person can capture relevant information or knowledge and will then decide whether to apply this/put it into practice. A person can also appraise a certain practice and decide not to apply it. Still, that person gained new knowledge as a result of the learning experience and has the freedom to decide whether to apply the new knowledge at any moment he or she wishes to do so. Exchange visits under SCI-SLM produced relevant new information for the participating community members. In a number of cases there is evidence that community members became inspired by an initiative and started to practise the specific technology or social organisation back home. Examples have been described in Box 9.1: these cases show tangible impacts of the exchange visits.

But, in a number of cases, the evidence is only implicit, when for example the learned skills are found to be useful but not immediately put into practice. It is of relevance when people themselves express their interest and or appreciation. In such cases it is important to observe the attitude of the community members. Both country teams from Ghana and Morocco have reported such observations. What is crucial is what communities themselves consider to be valuable knowledge or information. Numerous SCI-SLM community members learned about relevant technologies and reported back that they were inspired by the initiatives, even though they didn't put them into practice back home straight away. The Moroccan community members were impressed by the women of Moatani in Ghana, which might influence their thinking on gender issues back home, even though it might not immediately be translated into visible results of their awareness.

Cross-learning is an ongoing process and, of course, if learned skills are put into practice, this will often be outside the usual project span. To be able to fully evaluate the impact of the exchange visits within the SCI-SLM project, a post-project impact assessment should be conducted. And to ensure cross-learning does not disappear, it must either have been institutionalised or picked up by another project or programme. What will be the ultimate legacy of the exchange learning under SCI-SLM? This remains to be seen.

References

Critchley W. (2007), Working with Farmer Innovators – A practical guide. CTA.

Critchley W. and Tuijp, W. (2012), South-to-South solutions to sustainable land management. UNEP-GEF article.

Di Prima S., Critchley, W. and Tuijp W. (2013), Methodology for working with innovative communities – The SCI-SLM experience. UNEP-GEF policy brief.

Kojwang, O.H. (2013), Mid term review of the project SCI-SLM. Prepared for UNEP

SCI-SLM TAG backstopping reports. SCI-SLM Regional Steering Committee meeting reports. Available online at http://reseaumarpbf.org/spip.php?article32 (last accessed 19/5/16).

10 Contributing to global environmental benefits

Saa Dittoh, Maxwell Mudhara, Conrad A. Weobong,
Stephen Muwaya and Mohammed Mahdi

Introduction

Nature in its wisdom provided for humankind environments that were largely 'complete' and sustainable. It is human activity that has caused significant local, national and global environmental problems such as the massive loss of biodiversity, land degradation, climate change, ozone depletion and other negative impact on the environment. Major human activities that have caused considerable detrimental environmental impact include conventional agricultural intensification, deforestation and forest degradation, overgrazing and other activities. Global environmental problems affect nations in ways which most citizens and even governments hardly consider. Though agricultural intensification has for example provided food for growing populations, it has done so at high environmental cost and significant reductions in ecosystem services (Tilman 1999). Land degradation contributes to global climate change, loss of biodiversity and damage to shared international water bodies (Pagiola 1999), in addition to its negative effect on agricultural production. Deforestation and forest degradation also lead to loss of ecosystem services such as carbon storage, biodiversity conservation and water and food security (Bowler *et al.* 2010). Clemencon (2006) also points out that global environmental problems have the potential to undermine national security. The SCI-SLM project in all the four countries identified community initiatives that addressed several of these environmental concerns.

Global environmental benefits are those that the global community enjoys knowingly or unknowingly and directly or indirectly as a result of interventions that result in positive environmental impact. They are usually public goods arising from interventions that accrue to not only local and national communities but to global communities as well. They typically include biodiversity conservation and improvements, restoration of farm and grazing lands, watershed protection, reductions in greenhouse gas emissions to reduce climate change, carbon sequestration and others. A major outcome of the SCI-SLM project has been contributions to local, national and global environmental benefits in the short, medium and long terms, and aspects of these have been discussed in one way or another in other chapters. This chapter focuses on global environmental benefits of the SCI-SLM interventions. Table 10.1 is a summary of the community innovations identified and promoted by SCI-SLM in the four countries. The evidence that the innovations contribute to global environmental benefits is discussed in the next section under four main sub-sections: (i) soil fertility improvement; (ii) community forest management; (iii) crops and range lands rehabilitation, and (iv) livelihoods and poverty reduction with gender consideration. As will be observed, the sub-sections do not imply mutual exclusivity.

Table 10.1 Summary of community innovative initiatives

Country / community	Location	Social/technical initiative	Key features of the initiative
GHANA			
Tanchara	Lawra District, Upper West Region	Social and technical	Organic fertilizers. Fire belts, non-burning and non-cutting for biodiversity.
Kandiga-Atosali	Kassena-Nankanna District, Upper East Region	Technical	Organic fertilization using field composting taking advantage of rains.
Moatani	West Mamprusi District, Northern Region	Social and technical	Gender-based group cohesion for organic fertilization using compost pits.
Zorborgu	Tamale Metropolis, Northern Region	Social	Fire belts, non-burning and non-cutting for biodiversity. Application of crop residues, animal droppings for fertility and trees on dam banks for stabilization.
MOROCCO			
Anzi	Ouneine	Social	Carpenter coop, potentially leading to reduced pressure on forest.
Lamhalt	Ouneine	Social and technical	Land reclamation, 6 to 7 ha annually.
Afourigh	Ouneine	Social and technical	Water rights. Innovative irrigation management system.
Agouti	Äit Bougamaz	Social	Coops working on handicrafts, part of shared profits invested in tree planting.
SOUTH AFRICA			
Mtabamhlope	Northern Drakensberg	Social and technical	18 ha of wattle forest managed by Mhulungwini Tribal Authority.
New Reserve	Northern Drakensberg	Social and technical	45 ha of wattle forest managed by a community development committee.
Gudwini	Mzinyathi District	Social and technical	Indigenous forest managed by local structures and guards.
Amazizi	Bergville, uThukela District	Social	Monitoring of cattle by the owners to reduce theft using a new system – leads to improved range management.

Table 10.1 Continued.

Country/community	Location	Social/technical initiative	Key features of the initiative
UGANDA			
RECPA	Rwaha, Ntungamo District	Social	Tree planting on degraded slopes by a self-created environmental association.
NACIA	Nalukonge, Nakasongolo District	Technical	Rehabilitation of termite-affected grazing pastures by paddocking and stoloniferous grasses.
Banyakabungo Coop Society	Ntungamo District	Social	Communal grazing management on degraded pastures.
Bandera 2000	Kamuli District	Technical	Promotion of fruit trees for income generation through SLM practices.

Source: Ghana, Morocco, South Africa and Uganda SCI-SLM teams.

Evidence-based results of contributions of community SLM initiatives to GEBs

Soil fertility improvements

Areas with serious SLM challenges in Africa are mainly in the arid and semi-arid parts of the continent and that is why those parts were targeted for SCI-SLM interventions in the four countries. In most of those areas, continuous cropping (with limited soil management) and conventional agricultural intensification has led to soil exhaustion, soil erosion and looming desertification. The soil improvement innovations of the people in the four SCI-SLM sites in Ghana and three sites in Uganda (see Table 10.1 and Chapters 5 and 7) were aimed at improving soil fertility and/or increasing soil organic matter, as well as improving agrobiodiversity. The agricultural lands became degraded mainly due to the aforementioned activities and there has been considerable loss of soil carbon. Loss of soil carbon implies increased emission of greenhouse gases and that is a major contributing factor to climate change. Soil mining through agricultural intensification, overgrazing, burning of crop residue, bush burning and similar activities prevents nutrients being returned to the soil and thus reduces the build-up of soil organic matter, resulting in greenhouse gas emissions. These activities also reduce storage of carbon in the soil and thus reduce the ability of the soil to be a carbon sink. The negative global environmental impact of land degradation is thus quite serious, though difficult to measure quantitatively. Any interventions or measures that reduce these negative impacts contribute to global environmental benefits.

The organic fertilization through composting by the women of Moatani and the farmers of Kandiga-Atosali in Ghana contributes to the reduction of greenhouse gas emissions and global warming and thus is a major source for climate change mitigation. It is also known that organic fertilization helps to conserve biodiversity because non-use of chemicals encourages a natural balance within the ecosystem and helps prevent domination of particular species over the others.

The rehabilitation of grazing pastures by paddocking and stoloniferous grasses in Nalukonge, the improved management of communal grazing in Ntungamo District in Uganda, as well as the non-bush and crop residue burning measures of the people of Tanchara and Zorborgu in Ghana are also ways of reducing the loss of soil carbon and thus reducing greenhouse gas emissions and global warming. It would have been more appropriate if these environmental benefits in terms of build-up of carbon stocks were computed, but that is beyond the capabilities in terms of the resources of the SCI-SLM project. An aspect of it was, however, attempted in Zorborgu, one of the Ghana SCI-SLM sites. The methodology used, although not very robust, assessed carbon stocks above and below ground levels in burnt and non-burnt farmlands as well as in natural savanna woodland and a teak plantation (Weobong 2013). Table 10.2 gives the results obtained from the computation of soil organic carbon (SOC) levels under the different land uses. The table shows that the non-burnt farmland and the natural savanna woodland had considerably higher levels of SOC than the burnt farmland.

Table 10.2 Soil organic carbon levels (in tonnes per ha) under different land use types

Land use type	Minimum	Maximum	Mean	Std deviation
Burnt farmland	188.93	209.58	198.72	1.036
Non-burnt farmland	294.13	326.49	307.63	1.683
Natural savanna woodland	322.17	371.58	351.95	2.621
Teak plantation	213.68	264.15	238.62	2.524

Source: Weobong 2013.

Soil degradation also leads to the destruction of below-ground biodiversity; there are numerous soil organisms that help to maintain nutrient cycling, soil structure, moisture balance and the overall fertility and productivity of the soil (Pagiola 1999). The soil fertility improvement innovations in the various communities are instrumental in restoring or improving such below-ground biodiversity. Agricultural intensification through monocropping, overgrazing and bush burning also result in above-ground loss of biodiversity. Tanzubil *et al.* (2004) found that intensive rice production in the Bawku area of Ghana resulted in commercial rice farmers having no knowledge of several indigenous rice varieties. It is a great global loss to destroy fauna and flora. A wide variety of fauna and flora is a valuable source of genetic material for improvement of the tolerance of crops and livestock to drought and diseases (Hassan and Dregne 1997). The non-burning and general land restoration goals of the community innovations help maintain valuable fauna and flora and this contributes valuable global environmental benefits.

Community forest management

As indicated in Table 10.1, several of the community initiatives identified for the SCI-SLM project are community forest conservation and management innovations. Sustainable forest management (SFM) has been of great importance at local, national and global levels because of its economic and environmental benefits. SFM aims at maintaining and enhancing the economic, social and environmental value of forests for the benefit of present and future generations (Bowler *et al*. 2010). It has been shown in many situations that SFM is best obtained through the involvement of forest communities; hence the importance of community forest management. The main global environmental benefits of forests include biodiversity conservation and carbon storage. There is an increase in carbon sequestration when forests are prevented from being destroyed. The trees trap carbon, leaving less carbon dioxide in the atmosphere, and release oxygen into the atmosphere. They help to increase above-ground carbon storage. That is why carbon sequestration projects in several countries under the Kyoto Protocol aim at encouraging communities to plant more trees.

Three of the four community initiatives in the South Africa SCI-SLM project consisted of different forest management systems. Though the motivation for the initiatives had little to do with biodiversity or carbon sequestration, the effect and eventual impact, with SCI-SLM intervention, is global environmental benefits of increased biodiversity and carbon storage. In KwaSobabili and Reserve B (New Reserve), both in the uThukela District of KwaZulu-Natal in South Africa, 65 ha and 40 ha, respectively of wattle trees have been protected and managed by the communities since 2000 and 1992, respectively. The SCI-SLM project has encouraged them by assisting them to obtain a forest conservation licence for the invasive wattle trees and with good ideas of community forest management. The third community initiative is conservation of a 45 ha indigenous forest by Gudwini community in the uMzinyathi District since 1945. The initiative is spearheaded by the Chief, and only dead trees are allowed to be cut for firewood. That is clear indication that there has been considerable carbon sequestration by that forest for decades, and that is being improved by the SCI-SLM intervention. The community members regard their biodiverse indigenous forest as an asset for fruits, medicinal plants and fuelwood. An ecosystem service that has been under threat as a result of consistent loss of biodiversity is availability of medicinal plants. In Ghana, one of the reasons given by the Zorborgu community members for the protection of their community forest is the provision of medicinal plants. As indicated in Chapter 5, the Chief of Zorborgu claims that 'there are herbs in [their] forest to cure every disease'. Biodiversity has local and national benefits but it also has significant global benefits. Fauna and flora are important genetic and pharmaceutical resources of global importance. Good management of forests can also allow sustainable harvesting of fuelwood to avoid reductions in carbon sequestration.

Other examples of SCI-SLM interventions with high potential for carbon seques-tration are the Anzi community carpentry cooperative and the Agouti handicraft cooperative initiatives in the Atlas Mountains of Morocco. The Morocco SCI-SLM encouraged the two community cooperatives to out-scale their social innovations as they have the potential to reduce pressure on the forest, in the case of the former, and create forest, in the case of the latter, and thus increase carbon sequestration. Even the Bandara 2000 community initiative in Uganda which aimed at promoting fruit trees for income has carbon sequestration potential. According to Reicosky (2003), agricultural carbon sequestration may be one of the most cost-effective ways to slow the processes of global warming. Global warming is probably the greatest threat to environment security, food security and the health of humankind.

Crop and range lands rehabilitation

Land degradation encompasses several destructive phenomena including soil erosion, desertification, loss of soil fertility, salinisation and others. Thus, aspects of land degradation have been dealt with already. The emphasis here is on physical land degradation such as the removal of tree and grass cover and soil erosion and the like on crop and range lands. According to the Global Environment Facility (GEF 2014), land degradation is a major threat to biodiversity, ecosystem stability and society's ability to function. Loss of biomass through removal of vegetative cover and soil erosion releases carbon dioxide and at the same time reduces the carbon seques- tration potential of the degraded land. This in turn results in the production of greenhouse gases which contribute to global warming and climate change. Climate change may in turn increase land degradation through soil erosion, droughts, flooding, increased wind speed and several other factors. According to the United Nations Convention to Combat Desertification, desertification is land degradation in the arid, semi-arid and dry sub-humid areas resulting from various factors, including climatic variations and human activities (WMO 2005). It is predicted, through climate change modelling, that, with a doubling of atmospheric carbon dioxide content, there will be a 17 per cent increase in desert land worldwide due to climate change (WMO 2005). Thus, the community SLM initiatives against land degradation are actually actions against desertification.

The Lamhalt community initiative in the Ouneine Province of Morocco involved reclaiming 6 to 7 ha of degraded land annually (see Chapter 6), and the Amazizi community initiative in the uThukela District of South Africa involved monitoring cattle movements leading to improved range management. In Uganda, the RECPA in Rwaha in the Ntungamo District rehabilitated degraded slopes by planting trees; the NACIA in Nalukonge in the Nakasongolo District rehabilitated grazing pastures by paddocking; and the Banyakabungo Cooperative Society in the Ntungamo District rehabilitated degraded pastures through communal grazing management. All these activities are very important actions against desertification, global warming and climate change. Sustainable and diversified land management practices are the most effective ways of avoiding land degradation and rehabilitating degraded lands.

Water degradation through pollution, silting of rivers, over-exploitation of groundwater, salinisation, droughts and floods are serious problems of global concern. Climate change aggravates the problem. The Afourigh community initiative in the Ouneine Province of Morocco tackled the problem of water rights and irrigation water management (Chapter 6). The Zorborgu community initiative in Ghana also protected their dam by protecting their community forest and by planting grasses and trees on the banks of the dug-out. The organic fertilization initiatives of Moatani women and Kandiga-Atosali in Ghana are measures that reduce the use of inorganic fertilizers which have significant water-polluting and even soil degradation potential. All these community activities are generally intended to protect and stabilise the ecosystems and thus contribute to reduction in climate change.

Livelihoods, gender and poverty reduction

All the community initiatives discussed above had a central aim of improving the livelihoods of the community members and reducing poverty in their difficult environments. As stated in Chapter 5, 'the main driving force behind all the innovations is the desire of the people to effectively, efficiently and sustainably improve the management of their farm and forest lands in order to improve their livelihoods and to maintain a healthy environment'. Although this was stated with respect to Ghana, it is the case for all four countries. Improvements in livelihoods and poverty reduction may seem to be local and national benefits, but they have very significant global benefits of food, health and environmental security. That is why, to meet the Millennium Development Goals, countries pledged to ensure that policies designed to conserve internationally important ecosystem services in forests fully take account of impacts on local livelihoods (Bowler *et al.* 2010) and that is also why the first MDG aimed at poverty and hunger reduction and the first two Sustainable Development Goals (SDGs) aim at eradicating both poverty and hunger. As discussed above, all the community SLM initiatives result in various global environmental benefits, particularly the reduction in greenhouse emissions and increase in carbon sequestration. They all, however, also have the potential to improve livelihoods and reduce poverty of local land users as well as increase global food security. All the community initiatives also recognised the importance of gender roles and involved all genders effectively in the activities. Indeed the Moatani (Ghana) women's collective action in organic fertilization impressed Kandiga-Atosali (Ghana) women and men farmers, Moroccan visiting farmers and the SCI-SLM team to think of forming and promoting men's and women's farmer groups (Chapter 9).

References

Bowler, D., Buyung-Ali, L., Healey, J.R., Jones, J.P.G., Knight, T. and Pullin, A.S. (2010), The evidence base for community forest management as a mechanism for supplying global environmental benefits and improving local welfare. *CEE Review* 08-011 (SR48). *Environmental Evidence*. Available online at www.environmentalevidence.org/SR48.html (last accessed 19/5/16).

Clemencon, R. (2006), The costs and benefits of protecting global environmental public goods. *Global Commons*, Chapter 2, pp. 31–742.

GEF, (2014), Available online at www.thegef.org/gef/land_degradation (accessed on 4/09/2014).

Hassan, H. and Dregne, H.E. (1997), Natural habitats and ecosystems management in drylands: An overview. Environment Paper, 51. Washington. DC, World Bank.

Pagiola, S. (1999), The global environmental benefits of land degradation control on agricultural land. World Bank Environment Papers Series, 16. Washington, DC, World Bank.

Reicosky, D.C. (2003), Conservation agriculture: Global environmental benefits of soil carbon management. *Conservation Agriculture*, pp. 3–12.

Tanzubil, P.B., Dittoh, J.S. and Kranjac-Berisavljecvic, G. (2004), Conservation of indigenous rice varieties by women of Gore in the Northern Savanna Zone, Ghana. In Gyasi, E.A., G. Kranjac-Berisavljecvic, E.T. Blay and W. Oduro (eds) *Managing Agrodiversity the Traditional Way: Lessons from West Africa in sustainable use of biodiversity and related natural Resources.* United Nations University Press, pp. 97–105.

Tilman, D. (1999), Global environmental impacts of agricultural expansion: The need for sustainable and efficient practices. *Proc. Natl. Acad. Sci. USA* 96, pp. 5995–6000.

Weobong, C. (2013), *Increase in Total System Carbon in Ghana*. Faculty of Renewable Natural Resources, University for Development Studies, Tamale, Ghana.

WMO, (2005), *Climate and Land Degradation*. WMO, 989. World Meteorological Organization (ISBN 92-63-10989-3).

11 SCI-SLM methodology

Refinement of the original design

Sabina Di Prima, William Critchley and Eva van de Ven[1]

The original design and hypothesis underpinning the SCI-SLM methodology was field tested during the four years of the project. A further refined methodology for upscaling and institutionally embedding SLM initiatives was planned to be one output of the SCI-SLM project itself. This chapter summarises experience with the methodology at both national and regional levels, and makes suggestions for its further evolution and improvement so that it can be applied more widely in initiatives of this nature.

Working with a flexible methodology

The refinement and further development of the methodology was set as a component and output of the SCI-SLM project. In practice, this required a 'learning-by-doing' approach to allow the methodology to evolve in response to local conditions and institutional settings. SCI-SLM's design was based on the project's 'results framework' (UNEP 2009:30) but encouraged tailored-made solutions, decisions and approaches. Therefore, at the outset of the project, it was not the intention to provide one single blueprint methodology for all countries to work with – but to allow each to develop their own from a set of basic principles. Throughout the project, two main mechanisms were used to steer and assess the refinement of the overall methodology: regular backstopping missions/field visits by the technical advisory group (TAG), together with the national teams (23 missions in four years) and the annual Regional Steering Committee meetings, held in turn in each of the four participating countries (South Africa in 2009, Morocco in 2010, Uganda in 2011 and Ghana in 2012). At the end of the project period, the methodology refinements have been assessed and evaluated by the national teams. Information was gathered from each country through the use of a survey (SCI-SLM 2013), and this information was supplemented by the TAG's experience with the methodology. Herewith, we present how the original methodology design has developed and reflect on what the common denominators are, within the multiple approaches used. As it will become clear in this chapter, there are still outstanding issues regarding refinement of the methodology.

A regional approach

The participating SCI-SLM countries, located in the four corners of Africa, allowed the methodology to be tested in diverse socio-economic circumstances and distinctive biophysical environments while dealing with common problems of land degradation. Each country approached the programme in its own way. Not only did the national teams adjust the methodology and guidelines to their specific context and needs, but they also chose specific thematic or disciplinary emphases so as to focus on their priority areas.

Participation of four countries also allowed for diverse institutional players and settings – government, NGOs, universities, private organisations, parastatals, farmers' organisations and local associations worked side by side in this project. Lead agencies in each country had been chosen on the basis of their comparative competence and experience in similar situations, working with farmer groups. Taking into account the diversity of the lead agencies in each country (see Table 11.1), with their accompanying frameworks and approaches, the countries all had different starting points. The supervisory agencies acted as bonding elements and points of reference in this organisational structure.

Table 11.1 SCI-SLM consortium

SCI-SLM lead agencies		
Country	*Agency*	*Type of agency*
Ghana	University for Development Studies (UDS)	University
Morocco	Targa-Aide	NGO
South Africa	Centre for Environment, Agriculture and Development, University of KwaZulu-Natal	University
Uganda	Ministry of Agriculture, Animal Industries and Fisheries (MAAIF)	Governmental

SCI-SLM supervisory agencies

Executing organisation
Centre for Environment, Agriculture and Development (CEAD), University of KwaZulu-Natal, South Africa

Technical advisory group
Centre for International Cooperation, Vrije Universiteit Amsterdam (CIS-VU), Netherlands

Implementing agency
The United Nations Environment Programme (UNEP), Kenya

Source: UNEP, (2009), Stimulating Community Initiatives in Sustainable Land Management (SCI-SLM). Project document.

The process of bringing together different players and contexts within the SCI-SLM project and assembling these under one basic methodological denominator presented a valuable opportunity for the different African countries to promote South-to-South learning, while testing the impact of stimulating local community initiatives on the ground. Ultimately, on a national basis, the SCI-SLM project had the effect of bringing together these various experiences and learning points, thereby sharpening the concept and mechanisms of such a bottom-up approach, and institutionalising this in the relevant government ministries and associated organisations.

National 'starting points'

Although all different, the methodologies that were already in place amongst the four lead agencies before the advent of SCI-SLM presented overlap through their common acknowledgement of the value of indigenous knowledge and local innovation amongst land-based communities. In the case of Ghana and Uganda, the lead agencies already had particular, well-defined approaches in place for working with farmer innovators and applying participatory methodology in sustainable land management. Their methodological starting points are presented below.

Ghana

The University for Development Studies (UDS) based in Tamale, northern Ghana, uses the 'Plug-in' as the guiding principle of their community innovation approach. The 'Plug-in' principle focuses on integrating indigenous knowledge and cultural practices of local communities with scientific knowledge (see Figure 11.1). The result is referred to as 'hybrid knowledge'. The 'Plug-in' principle has been developed to engage and help land-based communities to improve what they are already doing themselves. The main goal is to promote effective and sustainable development in the communities with the involvement of the key stakeholders. The principle is based on the premise that new knowledge and ideas from outside (external interventions), although important, can never replace what exists, namely what people know and have practised over time. External interventions are desirable only if they bring improvements to the existing situation. This implies a thorough understanding and appreciation of the existing situation before making any intervention. In line with the 'Plug-in' principle, UDS students live in communities during one semester to carry out specific assignments. This arrangement continued under SCI-SLM. Students helped specifically in the identification and characterisation of the community initiatives as well as other implementation aspects of the project. The 'Plug-in' principle married seamlessly with the SCI-SLM approach and helped in shaping it further with field application.

Uganda

In Uganda, Promoting Farmer Innovation[2] and PROLINNOVA were influential forerunners to the SCI-SLM project (see Chapter 2). As a result of strong governmental involvement in these projects and their record of successful experiences, the Ministry of Agriculture, Animal Industry and Fisheries decided in 2007 to kick-start the SCI-SLM country project on its own initiative and funding, before the overall SCI-SLM project officially started. The first national co-ordinator had been also the co-ordinator of the PFI programme in Uganda, and thus the step across to

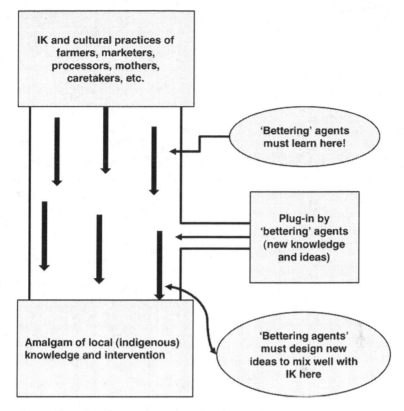

Figure 11.1   The UDS Plug-in principle (SCI-SLM 2009)

SCI-SLM with its similar methodology was an easy transition (see Chapter 3). In practice, these competitive advantages translated into a quick start-up of the project's field activities as well as the most extensive experience with the methodology amongst the four SCI-SLM countries.

Working with the SCI-SLM methodology design: an evaluation

As explained in Chapter 3 of this book, the SCI-SLM methodology is based on a basic and effective methodology previously developed under the Promoting Farmer Innovation project. This methodology, designed and tested with individual farmer innovators, was then modified for use with communities under SCI-SLM. The methodology consists of two components, 'programme development processes' for vertical scaling-up towards institutionalisation (sustaining the process) and 'field activities' for the improvement of community initiatives and horizontal spread (sustained management of innovation centred on communities).

This ex-post assessment of the methodology implementation and its refinements begins with the field activities and is broadly organised under three categories: (i) what worked well; (ii) what aspects of the methodology required adapting/tailoring to fit with the specific situations at country level; (iii) and what issues are still outstanding. The assessment itself is based on each country's perceptions of how the methodology worked in practice and the experience of the TAG team itself.

Field activities

The original ten-step methodology for field activities is thoroughly explained in Chapter 3. In brief, it guides the process of identification, selection, characterisation, analysis, improvement, documentation and dissemination of community initiatives through ten steps. The final goals are further stimulation, empowerment, improvement and spread of community initiatives. Its graphic outline is reported in Figure 11.2 for ease of reference. In this chapter, we will revisit the field activities with special emphasis on the identification of common denominators in terms of challenges encountered and solutions developed. In general, all countries made small adjustments in the field activity methodology. These adjustments were mostly derived from experiences and lessons learnt in the field while working in close contact with the innovative communities.

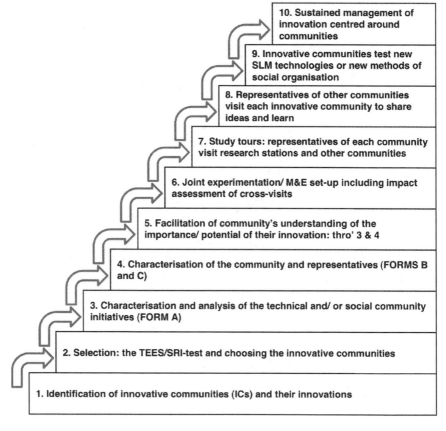

Figure 11.2 Field activity methodology – as designed for Stimulating Community Initiatives in Sustainable Land Management (Critchley and Di Prima 2009)

Step 1: Identification of innovative communities and their innovations

In the early phase of the project, looking for evidence of an innovative SLM practice initiated by a community or (even more so) of a novel method of social organisation proved to be a challenging task for the national teams, as they had to familiarise themselves with the SCI-SLM principles and approach. In practice, the teams had to find the identification strategy best suited to their own context and implementation capacity. Most teams started the identification of innovative communities by using informal networks and key informants. However, the means of identification were progressively refined, and time requirements simultaneously reduced, as teams became more familiar and confident with the methodology. The team from Morocco, for example, chose to use systematic surveys of agricultural production systems in combination with interviews of key informants (development agents, farmers, etc.) to identify community initiatives and simultaneously collect relevant background information on the agrarian and production systems of the *douars* (villages). In Ghana, a more comprehensive list of potential community initiatives was drawn up with the help of the SCI-SLM Steering Committee members and the stakeholders and partners who attended the official kick-off workshop of the project. The South African team approached identification by setting up multi-stakeholder meetings with representatives of government departments, universities and civil society organisations. They also made use of the relevant networks (e.g. PROLINNOVA) and platforms (e.g. Sivusimpilo-Okhahlamba Farmers' Forum) in which CEAD is an active member.

The identification step was less challenging for the Ugandan team, given the well-embedded experience with the similar PFI methodology. In Uganda, communities were identified mostly through government agricultural extension agents who work in close contact with rural communities. It should be noted that, in all countries, a wealth of information concerning the communities was gathered at the identification step but this was not always systematically documented. In general, for all countries, it proved an exciting experience to uncover innovative solutions in remote and marginalised rural communities. It was also noted that young people and women (in the cases of Morocco and Ghana respectively) had a strong role in establishing initiatives and bringing change within traditional social structures and gender constructs (see Chapters 5–6).

Step 2: Selection of innovative communities: TEES-test and SRI-test

The TEES-test and the SRI-test, in combination with the other selection criteria outlined in Chapter 3, were used by the national teams to gauge whether the identified SLM initiatives: (i) met the project requirements and (ii) had upscaling potential. In simple terms, the TEES-test and SRI-test provided the national teams with a more structured assessment framework to answer three basic questions:

1   Is this a good technical and/or social innovation in the field of SLM?
2   What is the relevance and potential of this initiative for other communities that face similar problems?
3   Is it worth investing in its further improvement and dissemination?

The TEES-test proved to be very useful and effective in the assessment of technical initiatives since the criteria (**T**echnical effectiveness, **E**nvironmental friendliness, **E**conomic validity and **S**ocial acceptability) were generally easy to explain and understand. In one country case, the criterion of 'social acceptability' was perceived to be redundant on the premise that a community initiative is, by default, accepted by its members. However, a 'community' rarely represents society at large, and hence the need to assess its acceptability from an outer perspective – the perception of those who are not part of/involved in the community initiative. It was also noted that the indicators of 'social acceptability' are not as easy to conceptualise and measure as those used in the TEES-test. In general the application of the TEES-test could have been further sharpened by complementing the skills of the national teams with the greater capacity and understanding of subject-matter specialists.

The SRI-test (**S**ustainability, **R**eplicability and **I**nclusiveness) for the assessment of social innovations was less well understood. One reason for this was that it had not been sufficiently thought through at the onset of the project. Having said that a good social innovation should ideally be sustainable, replicable and inclusive, the test fell particularly short in explaining the exact relevance and characteristics of this type of innovation in the field of SLM. Therefore, as part of the project's methodological approach, a support study was carried out in Uganda to better understand the particularities of social innovation and to formulate suggestions on how to further strengthen the SRI-test. The study was conducted by one of the co-authors of this book chapter (van de Ven 2011) and co-supervised by the TAG (Vrije Universiteit Amsterdam) in collaboration with members of the Uganda SCI-SLM team (MAAIF). While too late to influence the identification process which had already largely taken place, the findings of the support study are very pertinent and are summarised in Box 11.1.

Box 11.1 Social innovation in SLM – a case study from Uganda

Premise to the support study

SCI-SLM aimed at building better recognition of social (group) innovation within the rather technically focused field of SLM. The motivation for this being that it is crucial to address social challenges, including contextual themes such as land tenure, gender diversity, access to information, financial resources and health issues that hamper agricultural development goals. Acknowledging and analysing processes of social innovativeness amongst communities allow these 'intangible' processes to receive the recognition that they deserve and thereby help tackle land degradation problems in a holistic manner.

The support study in Uganda

By means of a participatory research design, two rural communities in Uganda were studied to find evidence of social innovation as well as to analyse the true potential and relevance of this category of farmer innovation for improved SLM.

The specific research objectives were: (i) to define the concept of social innovation; and (ii) to reassess the SCI-SLM criteria for a 'good' social innovation.

A more inclusive working definition of social innovation was developed: 'The process of creating or renewing systems of social order and cooperation which govern the behaviour of a set of individuals within a given human community, with the aim of improving agriculture and the environment and thus strengthening livelihoods' (van de Ven 2011).

The fieldwork generated strong evidence of social innovation in the two studied communities:

- *Banyakabungo*: this grazing land management society demonstrated how land could be owned and managed communally in order to safeguard tenure rights, while ensuring quality and productivity. Amongst the paid-up society members, knowledge and ideas about cattle and land management, and income from their produce was shared. Some members seemed to carry more responsibility than others, which translated into working more hours. Income was divided according to 'shares'. However, everyone had an equal interest in contributing to maintenance of the land and keeping the cattle in good health as continued and secure tenure of the land by the group was contingent on this. A democratic decision-making organ was in place and administration well documented.
- *BANDERA 2000*: this group's overall mission is to fight against poverty, vulnerability, sickness and hunger by improving agriculture and the environment, and to generate income by creating agricultural enterprises and connecting farmers to work and innovate together. The people in this group share knowledge about the latest land management practices amongst themselves and are stimulated to keep each other informed through monthly meetings, connecting with other farmer groups in the region and organising workshops in the demonstration gardens. An ambitious five-headed board led the association. Women were especially motivated to join the group since the board recognised that they do most of the work on the land and in the household.

On the basis of the collected and analysed evidence, some general conclusions were drawn:

- Social innovations do not exclusively aim to improve productivity and take better care of the land, although improving productivity and overcoming poverty are the most important drivers of social innovation.
- By working together and involving more people in contribution and benefits, the communities address exactly those social issues that hamper agricultural development as well as social and economic development in their broader community. These communities have developed new ideas to break through those social barriers. Issues such as land tenure, empowerment (of women and other vulnerable members of society), health and education are critical issues

when analysing social innovation related to SLM. Precisely these contextual social processes need to be dealt with and to be given attention in order to achieve sustainable land management.

The study recommended that a more relevant set of criteria is needed to decide whether a social innovation is 'good' and whether it is eligible for SCI-SLM, in place of the 'SRI-test' (Sustainability, Replicability and Inclusiveness). Based on the study and later reflection the key criteria were determined to be:

- *Sustainability* (endurance) of the social initiative
- *Efficiency* (economic viability) of the social initiative
- *Replicability* (its potential to spread to other communities)
- *Responsibility* (benefiting multiple people – and not harming society).

The 'SERR-test' was developed. Efficiency and responsibility were additions to the original SRI-test and inclusiveness was removed as a separate heading. The list could have been further extended with other social properties that stood out in the case studies, such as future vision, empowerment, leadership and democracy. In fact, behind each community innovation there is a social mission. Thus it is important to ask: In what way are people involved in group dynamics, is the innovation actually a group undertaking (benefitting > 2 people) and is there equity amongst the members in this mission?

Furthermore, the study put forward several recommendations to further integrate the concept of social innovation into SLM and within the farmer innovation methodological framework:

- to continue in-field research, establishing close collaboration between communities, researchers, local extension agents and local university students, to create hybrid knowledge and have new insights;
- to conduct research and interviews in the local language so that community members can clearly communicate and express themselves;
- to involve a local agricultural officer, where possible, to ensure monitoring and evaluation and follow-up after research has finished;
- SCI-SLM forms should be used to store collected data and share with the national team and partners, and to see whether cross-visits could be arranged according to the information that is described in the forms;
- more land-based communities should be visited by local agricultural (extension) officers and, where possible, students to observe efforts of social innovation;
- in order to upscale the farmer innovation methodology, it is important to keep acknowledging social innovations that are created alongside technical innovations, to prove social innovation's huge impact in the field of SLM and small societies' capacity to make positive changes.

Source: Based on *Communities Taking the Lead*, Eva van de Ven 2011.

The SCI-SLM team in Uganda decided to further strengthen this crucial method-ological step of selection by developing a complementary list of selection criteria with a scoring system (see Box 11.2 below). In their view, this additional tool was justified by the need to take into account some relevant aspects not fully encom-passed in the original selection toolbox. The specific aspects covered in the complementary list of selection criteria were: (i) the community's strength in data collection, monitoring and documentation; (ii) community cohesion, manage-ment rules, accountability mechanisms and democratic practices; (iii) level of community ownership; (iv) embedding of the community initiative into the broader social context and related opportunities for scaling-up; and (v) linkages between the innovative communities and locally established structures. It should be noted that this complementary list of selection criteria was tested and used only in Uganda, although its usefulness was recognised by the other SCI-SLM countries.

Box 11.2 Additional criteria for selection of local community initiatives in SLM (SCI-SLM Uganda)

The initiative should be selected from pilot sub-counties (minimum 1 and maximum 2 groups per sub-county) and the selection should be based on the criteria below:

Criteria

1. Organization and structure
(i)   Evidence of existence of organised and active community group for at least six months or more
(ii)   Strong community organisation encouraging participation of women and youth in all activities, i.e. leadership
(iii) The initiative formed by community with common interest, values, goals and identity
(iv)   Initiative formed by local community
(v)   Willingness to learn from and to share knowledge with others
(vi)   The initiative has been developed with little or no help/money from outside
(vii) Extent to which organisation is linked to local structures
       Subtotal (maximum score for this category):   5

2. Management/record keeping
(i)   Should have clearly defined constitution/guidelines
(ii)   Initiatives with in-built transparency and accountability mechanisms for the leaders
(iii) Evidence on democratic practices
       Subtotal (maximum score for this category):   5

3. SLM technologies (practices)
(i)   The initiative should pass the TEES-test
(ii)   Initiative that is technically and/ or socially innovative
(iii) Initiative is locally conceived or can be easily adapted to local conditions

(iv)   Initiative contributes to SLM and environment protection and conservation
(v)   Extent to which the whole value chain is addressed
(vi)   Initiative that direct benefits the community in terms of income, food security, etc.
(vii) Practices which have high chances for replication elsewhere (especially those demonstrating SLM)
(viii) Severity of the problem with relation to SLM
(ix)   Initiatives that promote indigenous knowledge and practices
(x)   Initiative which yields multiple benefits and which saves time and labour/energy for other activities

Subtotal (maximum score for this category):   **10**

4.   Enterprises and activities
(i)   Practices leading to diversification of household income and buffering against crop failure
(ii)   Availability of markets for the products
(iii) Practices that are simple, easy, cheap and requiring minimal specialised skills
(iv)   Practices with high potential for adoption and that allow for farmer-to-farmer learning
(v)   Practices which promote working through a bottom-up participatory process
(vi)   Initiatives with ability to leverage additional inputs from external sources and other projects
(vii) Initiatives that relieve pressure on the land resource (e.g. beekeeping, mushroom growing)

Subtotal (maximum score for this category):   **10**

Overall total   **30**

Finally, in Ghana, the University for Development Studies complemented the selection process with a list of additional criteria that built on and made more explicit specific aspects of the TEES-test and SRI-test. By way of example, some of the characteristics that were taken into consideration in the selection of the initiatives are reported below:

- *The number of years the initiative has been practised in the community* – it is hypothesised that the longer the period, the more likely will be its sustainability;
- *The number of community members taking part* – this is to some degree an indication that the initiative is popular through its high technical effectiveness and/or economic validity;
- *How widespread the practice is* – in terms of the number of communities in and around the area practising the innovation. If many communities are practising the initiative it is also an indication of greater technical effectiveness and/or economical validity. In this case, the original innovative community was sought out (see also next point);

- *Evidence of community ownership of the initiative* – it is important to verify that the initiative has its origin in that community and that it is embraced by community members;
- *Ease of horizontal and vertical upscaling* – a good initiative is one that can spread and benefit other communities; thus its ease of spread as well as its potential for further improvement are important characteristics to consider;
- *The (potential) impact on the livelihoods of the people* – an SLM initiative is most relevant if it has high potential to improve the social, environmental and living conditions of the people.

Steps 3 and 4: Characterisation and analysis of the technical and/or social initiative, the community and its representatives

The use of characterisation forms for each innovation, the innovative community and its key community members was a methodological achievement as these forms were able to capture important data (qualitative and quantitative) and simultaneously act as repositories of baseline information for later reference and analysis (see example in Box 11.3). Furthermore, completion and submission of these forms constituted a clear 'milestone' in country activities. After the first round of characterisation, the original forms were slightly modified by the technical advisory group (TAG) to make them clearer and capture additional baseline information. In particular, an eligibility checklist and a descriptive summary were added to form A (see Annex of Chapter 3, pages 37–40). Data collection proved to be a very time consuming and complex task. In some cases, data gathering required the observation and study of the community initiatives over a long period of time, often with the help of students living with the communities, so as to understand and fully appreciate the communities' dynamic and processes. Generally, it was easier for the national teams to capture qualitative rather than quantitative data. The latter were often partial information obtained from community members (especially the leaders) and may not always have been representative of the community as a whole.

The characterisation of those technical aspects leading to the measurement of global environmental benefits (such as net primary productivity; soil carbon stock; abundance, richness and heterogeneity of above- and below-ground biodiversity) were only partially addressed because of time and capacity constraints. For example, the SCI-SLM Ghana team started an inventory of the Zorborgu community forest to document the vegetation, the size and composition of plant species and the uses of the species by the people, but the study was not comprehensive (see Chapter 10). Furthermore, throughout the project, it became clear that more detailed guidance to the 'proxy indicators' of measuring GEBs as well as specific skills and capacity in the national teams were needed to carry out this type of assessment.

In general, the process of characterisation was appreciated and functioned well even though it proved more difficult and slower than had been anticipated. The entire SCI-SLM team perceived that community organisations were visibly strengthened in organisational and operational terms after participating in the initial three steps of the methodology, culminating in the full-fledged characterisation of their initiatives. The characterisation and validation process also strengthened the communities' understanding of the link and contribution of their innovations to sustainable land management. One outstanding issue remains as to how best to build on the characterisation forms for future monitoring purposes.

Box 11.3 Baseline analysis: summary of preliminary findings

Premise to the baseline analysis

The analysis of key aspects summarised below was derived from characterising 16 community initiatives (four per country). It was presented by the TAG team at the third Steering Committee workshop held in Uganda in 2011 with the objective of learning from and further refining the baseline data recorded in the characterisation forms. The database utilised a Microsoft Excel spreadsheet. Data categorisation and clustering were used for analytical purposes.

Summary of preliminary findings

- **Initiative typology:** identified six social initiatives, four technical initiatives and six initiatives combining both social and technical innovations. The TAG team remarked that it was a balanced mixture of initiative typologies from which to derive useful lessons.
- **Technical initiatives:** four in total of which two were from Uganda, one from Ghana and one from Morocco. The SLM technologies associated with these initiatives were: grazing land management (two), composting (one) and land rehabilitation (one). The TAG team noted that initiatives in wetland management and renewable energy, which seemed very relevant at the project proposal development stage, were not uncovered.
- **Social initiatives:** out of the six identified initiatives, Morocco, South Africa and Uganda had two each. The SLM technologies associated with these initiatives were: forest management (three), water management (one), grazing management (one), in-field SLM practices (one);
- **Combined social and technical initiatives:** six in total of which three were from Ghana, two from South Africa and one from Morocco. The SLM technologies associated with these initiatives were: forest management (four), composting (one), non-burning of community forest and fields (one). The TAG team noted that the category forest management included all technologies related to trees (on-farm and outside);
- **Area under SLM:** the total area under the initiatives was 1,840 ha based on information from 11 communities. Five community initiatives had no information on area. It was noted that one community initiative made up more than half of the total area;
- **Typology of innovative communities:** a wide and very diverse range, namely common interest groups (eight), whole villages (three), family group (one), women's group (one), working group (one), community development committee (one) and community association (one);

- **Size of community:** six communities had between 20 and 65 members, two communities had between 95 and 110 members, three communities between 250 and 500, one had 1 500 members. This information was missing for four communities and had to be collected;
- **Male to Female ratio:** three communities had more male members than females, another three communities had more female members than males, and two had equal numbers of male and female members. Eight communities had their gender composition unspecified/missing.
- **Starting dates of the initiatives** (organised per decade): three initiatives started respectively in 1945, 1970 and 1980. Seven initiatives started in the 1990s and six in the 2000s;
- **Source/origin of the initiatives:** eight were started by individuals (respectively three by village chief, one by village official, one by a village member and three not specified), five were started by a core group (of which two comprised young people), one by a family, two were not specified. The TAG raised the question of whether an initiative that started in 1945 really qualified as an initiative or a tradition;
- **SRI-test (Sustainability, Replicability and Inclusiveness):** of the six social initiatives, three passed the SRI-test in full. One only partly passed the test with sustainability in doubt [S]RI. One only partly passed the test with inclusiveness in doubt SR[I]. The last initiative had missing/unclear information for a selection decision to be made. The fact that some initiatives only 'partly passed' the SRI-test showed that improvements are needed before the initiative can be spread.
- **TEES–Test (Technically effective, Economically valid, Environmentally friendly, Socially acceptable):** of the four technical initiatives, two passed the TEES-test in full. One partly passed the test with environmental friendliness, and social acceptability in doubt TE[E][S]. One partly passed the test with social acceptability in doubt TEE[S]. The fact that some initiatives only 'partly passed' the TEES-test showed that improvements are needed before dissemination;
- **SRI-test and TEES-test of the combined social and technical initiatives:** of the six initiatives in this category, three passed the SRI and TEES tests in full. Two passed the SRI test, but did not have enough information regarding the TEES-test to make a proper assessment. One partly passed the tests (S[R]I and TE[E]S) with doubts in several aspects. The initiatives that only 'partly passed' the tests require improvements before dissemination can take place;
- **Monitoring:** eight of the identified communities had no monitoring system, four carried out monitoring, and two had partial forms of monitoring. Information on monitoring was missing for two communities. The TAG team noted that M&E was urgently needed for capturing data on indicators.
- **Documentation:** six of the identified communities did not have documentation, five had basic documentation, one community had developed a web site, while four had been documented through students' studies;

- **Adoption of initiatives by other communities:** A preliminary analysis of baseline data – a 'stocktaking' exercise - which took place some 24 months after the project had become active, showed that of the 16 SCI-SLM community initiatives: five CIs had not been adopted by other communities; four CIs had been adopted by either one or two other communities; five CIs had been adopted by three to five other communities; and two CIs had been adopted by ten or more communities (see Table 11.2). However, it was noted that adoption could not, in all cases, be entirely and unequivocally attributed to the SCI-SLM project. Other factors, such as the interventions by other (national) projects or the extension services, or indeed spontaneous community-to-community spread may have played a role in the adoption.

Source: Di Prima S. and Critchley W. 2011. Preliminary Findings from Baseline Analysis. SCI-SLM Working Paper.

Step 5: *Community members are facilitated in understanding the importance and potential of their innovation and helped to be able to explain this to others*

This step was intended to increase communities' confidence in their own initiatives with the dual goal of enhancing (i) the long-term sustainability of the initiatives; and (ii) the ability of community members to share their experience and know-how with others. In practice, it took considerable effort to make the communities realise that their sustainable land management initiatives were 'scientific' (TEES-test and/or SRI-test compliant) and, in some cases, needed only some minor improvements. In Ghana, although to a certain degree also in the other countries, there was a common perception amongst communities that their initiatives were not good enough and that they needed more 'modern' technologies (Kojwang 2013). There was also the case of the Amavimbela group in South Africa who had not even realised that their social initiative (monitoring of cattle movement and recovery of stolen animals) had an extraordinarily positive side effect on grazing land management.

Generally, countries felt that the facilitation of community understanding should be given more emphasis as this step is crucial for the successful stimulation of the initiatives and their upscaling. Given the opportunity and time, communities can clearly articulate their ideas, explain their initiatives and account for their resources. It was also noted that, when women are offered equal opportunities with men, they can be exceptionally responsive despite some initial intrinsic constraints (e.g. widespread illiteracy rate among women). However, the bottom line is that not all steps of the methodology can be allocated as much time as might be desired: in such a programme time carries an opportunity cost and must be planned judiciously. Therefore, it is inevitable to focus on a limited number of valuable community initiatives with high potential for spread in order to give them adequate attention and visibility. By way of example, in Ghana, researchers from the University for Development Studies, extension agents from the Ministry of Food and Agriculture

Table 11.2 Adoption of SCI-SLM community initiatives by other communities (after 24 months from the beginning of the project)

Country	Name of community	Initiative typology: Social (S) and/ or Technical (T)	SLM main technology category	Adoption
Ghana	Zorborgu	S&T	CFM – non-burning of community forest and farms	0
Morocco	Anzi	S	CFM – forest protection	0
Morocco	Afourigh	S	Water management	0
South Africa	New Reserve B	S&T	CFM – productive management of invasive species: wattle forest	0
South Africa	Gudwini	S	CFM – forest protection	0
Ghana	Tanchara	S&T	CFM – non-burning/ non-cutting of community forest	1
Ghana	Kandiga	T	SFM – non-burning of crop residues on farm and composting	1
South Africa	Ntabamhlophe	S&T	CFM	1
Ghana	Moatani	S&T	SFM	2
Morocco	Agouti	S&T	CFM - forest protection	3
Uganda	NACIA	T	Grazing land management	3
Uganda	Banyakabungo	T	Grazing land management	4
Uganda	Bandera 2000	S	In-field SLM practices for horticulture	4
Uganda	RECPA	S	CFM-reafforestation	5
South Africa	Amavimbela	S	Grazing land management	10
Morocco	Lamhalt	T	Land rehabilitation	17

Source: Di Prima, S. and Critchley, W. (2011), Preliminary findings from baseline analysis. SCI-SLM working paper.

Notes: CFM = Community forest management. SFM = Soil fertility management.

(MOFA) and representatives of several NGOs worked together with community members to understand the scientific, economic and social importance of the selected initiatives. Various constraints were also identified and addressed through dialogue and discussions. This close collaboration has undoubtedly been beneficial to all parties involved (community members, researchers and extension agents) in terms of greater belief in the communities' ability to be innovative and solve their own problems. Along the same lines, in Uganda it was perceived that regular visits to communities by the technical advisory group and the SCI-SLM team (both national and district levels) motivated and inspired the groups to improve their initiatives and scale-up their activities. This confirms the need to devote an adequate share of time and resources to this methodological step. However, the SCI-SLM team capitalised on its investment by entrusting the empowered communities as ambassadors of their message.

Finally, the country teams observed that, in general, understanding the importance and potential of the innovation is not always enough to propel scaling-up. In Uganda, for example, it proved difficult to spread social innovations in other contexts, because of their dependence on the cultural and socio-economic characteristics of the innovative communities. Upscaling would also require a set of additional conditions: improving group dynamics; strengthening monitoring and record keeping; building partnerships and negotiation skills; and mobilising resources.

Step 6: Joint experimentation and M&E

Joint experimentation was generally weak in all countries with the exception of Uganda, where two experiments (use of arboreal termites to control terrestrial termites and the use of a night *kraaling* system to rehabilitate degraded rangelands) were successfully conducted on the NACIA initiative in collaboration with Makerere University, the National Agriculture Research Organisation (NARO), the National Agricultural Research Laboratories Institute (NARLI) and the Ministry of Agriculture, Animal Industry and Fisheries. As explained in Chapter 8, a study visit to Kamwenge district was organised by the SCI-SLM team for the NACIA representatives. This visit, held prior to the set-up of the arboreal termites experiment, allowed the NACIA community to learn from the successful exper- ience in Kamwenge thus pre-empting some of the pitfalls in the joint experiment and ensuring effective participation of the NACIA community members in monitoring the establishment of arboreal termites. In Uganda, joint experimentation also took place in the Bandera community and concerned conservation agriculture. This was identified as a potentially interesting practice to experiment with by the chairperson of the Bandera community during a skill-sharpening workshop held by the SCI-SLM team in Kampala, Uganda. The group decided to experiment with this practice and in August 2011 established a demonstration plot in Nalimawa village with the assistance of the SCI-SLM team. Later the same year, the Bandera community members visited farmers in Pallisa district to learn more about conservation agriculture. The acquired knowledge was then integrated into the experimentation process. By April 2012, encouraged by the successful results of the first experiments, the Bandera community set up 30 conservation agriculture demonstration plots in Nawanyago and Kisozi sub-counties.

In general, the field testing showed that joint experimentation (involving community members and outside researchers) should be more thoroughly designed, planned and implemented. In addition, adequate resources (time, funds and people with technical expertise) should be allocated. Ghana was not unique in reporting that joint experimentation was not given enough attention largely due to time and resource constraints. Neither were joint experiments conducted in South Africa due to organisational changes and staff turnover. Clearly there are plenty of outstanding issues associated with joint experimentation built on indigenous knowledge. It is important to note that the ISWC II programme, referred to in

Chapter 2, made little headway with this aspect either (joint experimentation was not a feature of PFI). Joint experimentation, in combination with ad hoc support by formal/informal extension services and capacity building between innovative communities, could play a crucial role in the qualitative improvement of the initiatives. Thus, it is advisable that future programmes, if they wish to integrate joint experimentation, give specific emphasis and sufficient resources to it.

Monitoring and evaluation was also generally weak. In the project methodological guidelines, M&E was purposely non-prescriptive in order to give each country team the flexibility to tailor and set it up according to their needs and capacity. In practice, not all countries managed to put in place a comprehensive and systematic recording of data to capture relevant changes from the baseline (characterisation forms). It was also noted that communities' involvement in M&E was often not adequate. Communities should been provided with appropriate tools for data collection and simple analysis (e.g. soil-testing kits as piloted by the SCI-SLM team in Uganda), training and more accurate information to support their decision making. Also, as noted by the Ghana team, the comparison of M&E data between similar initiatives was only indicative as the communities operate in different physical and socio-economic environments, have different farming systems (including types of crops cultivated) and use different technologies with widely varying costs. However, some type of comparative evaluations can still be useful in identifying and promoting SLM practices which generate both short-term and long-term benefits, without significant trade-offs (see Chapter 5).

In principle, M&E should be strengthened and perceived as a valuable investment towards improved results and positive impacts. Also, it should not be singled out and attached to a specific step of the methodology but should be better visualised as a cross-cutting component running through the ten steps of the field activity methodology.

Step 7: Exchange visits/study tours

As explained in Chapter 9, SCI-SLM created a powerful learning platform for the selected communities through exchanges at various levels:

* visits between innovative communities within a country;
* bilateral visits between innovative communities of two countries;
* regional visits involving members of the innovative communities from all four participating countries; and
* study visits to research stations and other relevant (project) sites.

It was noted that this approach, based on mutual learning and sharing, was particularly effective and inspirational in the case of the most remote and marginalised communities, which had never had this type of exchange opportunities before. The exchanges, while explicitly aimed at enhancing the potential of the selected communities, proved to be a very valuable learning platform also for the numerous stakeholders related to the communities as well as the overall SCI-SLM project team. It is important to remark that farmers, researchers, extension officers and the other stakeholders involved interacted quite informally but very effectively during such visits.

Exchange visits allowed constructive peer-to-peer (farmer-to-farmer) learning, which in turn contributed to the qualitative improvement of some of the initiatives and the stimulation of new ideas/concepts/principles to try out. Technical innovations, in particular, showed higher potential to be replicated or adapted from one community to another, either within the same district but also across country borders. For example, the Moatani and Kandiga-Atosali communities in Ghana were approximately 80 km apart from each other and yet none of the two communities had knowledge of the fertility management technology of the other even though their basic problem (soil infertility) was the same. They only got to know of each other's technologies during the exchange visits. This widened their innovative horizons. New initiatives grew out of their expanded knowledge of what was possible, and new relations were established between communities. Another example concerns some of the water harvesting and management techniques observed in Morocco and Uganda, which were considered for experimentation at pilot level in Ghana (see Chapters 5 and 9).

Exchange visits and study tours also had the important side effect of enhancing appreciation by community members themselves of the worth of their own innovations. This created a sense of reassurance, provided encouragement and strengthened their commitment at the same time. In addition, exchange visits and study tours provided the opportunity to learn how to manage the initiatives sustainably. For example, the study tour organised by University of KwaZulu-Natal to the Natal Timber Extract in Greytown helped the KwaSobabili community learn specific technical aspects of wattle forest management to convert it from a wild, uncontrolled forest into a man-made, regulated forest, thus safeguarding their water resources from the invasive species (see Chapter 7). Furthermore, during the exchange visits, community representatives not only shared knowledge of their technical and social innovations but also compared (local) rules and regulations governing their initiatives.

Among the various forms of exchanges facilitated by SCI-SLM, the regional visits involving members of the innovative communities from all four participating countries received special praise as they helped break down some 'stereotypical' barriers to communication (language, culture, gender, perceptions and physical environment). The communities themselves made sensible choices on who should represent them in the regional visits. The representatives (men and women alike) were not always the leaders, but in general the most active members who were willing to travel and had a passport.

Step 8: Dissemination

This was a straightforward and natural next step in the SCI-SLM methodological sequence. Once the communities felt more grounded and confident of the value of their own initiatives (empowerment and validation process), they had pride and eagerness to invite representatives of other communities (outside the project) to visit and learn from their experience. Timewise, the first dissemination activities generally overlapped with the last round of exchange visits between innovative communities facilitated by the project. This confirms that dissemination can take place only after an initiative has reached a certain level of maturity and tangible results; the timeline will be case specific.

The organisation of field days at the location of some of the innovative communities (e.g. KwaSobabili community in South Africa and Bandera community in Uganda) with the involvement of neighbouring communities but also representatives of the local governments and the local traditional authorities played an important role in promoting horizontal spread. Another related example is that of the Bandera community in Uganda, which in June 2012 hosted the national cerebrations for the World Day to Combat Desertification presided over by the Minister of State for Agriculture, Animal Industry and Fisheries. During this event, several SLM technologies were showcased and demonstrated, while at the same time greater visibility was given to the SCI-SLM initiatives (beyond technologies).

In Uganda, the SCI-SLM team also helped the NACIA community in identifying partners to enhance the horizontal spread of the tested techniques for rehabilitation of degraded rangelands. In particular, the NACIA community used one of the planning meetings facilitated by SCI-SLM to develop, together with its partners (including NARO, MUK, MAAIF, NEMA and District Local Governments), a proposal to facilitate the horizontal spread of the arboreal termites and night *kraaling* techniques. The proposal for a small grant was submitted for funding to the UNDP/MAAIF SLM Enabling Environment Project to rehabilitate ten more degraded rangeland sites in Nalukonge community (see Chapter 8).

Dissemination was one of the strengths of the PFI programme (see Chapters 2 and 3) and it appears to have worked well under SCI-SLM too, although there are only preliminary data on impact assessments to confirm this impression. Most of the findings derive from Uganda, which is the country that ran the SCI-SLM methodology in full and for the longest period (officially from 2007, but some field activities had already taken place from 2004).

Step 9: Development and testing of new techniques and/or forms of social organisation

In Ghana, two of the selected communities realised that, with the help of the SCI-SLM team, the technical and social aspects of innovations could be complementary and strengthen each other. Thus, in the Moatani community, which had originally developed a social innovation, the SCI-SLM team encouraged improvement of the initiative's technical aspects. Conversely, in the Kandiga-Atosali community, with a technical innovation, the team encouraged community members to develop their social organisation (see Chapter 5). The two communities inspired each other and, in doing so, confirmed the importance of the role model potential of innovative communities.

In Uganda, a number of interesting innovations stimulated by exchange visits required capacity-building support for the visiting communities. In this respect, the SCI-SLM team made a concerted effort to ensure direct 'community-to-community' capacity building in order to enhance adoption and scaling-up. For example, following a visit to NACIA by the Bandera community in 2011, SCI-SLM created the conditions so that a NACIA representative could train members of the Bandera community in constructing an underground tarpaulin-lined water harvesting tank. Since then, the Bandera community has constructed 16 such water harvesting tanks on their own. Furthermore, the Bandera community hosted a training-of-trainers workshop on conservation agriculture for members from six Cattle Corridor districts in Nalimawa and Izanhyiro villages (see Chapter 8).

Step 10: Sustained management of innovations

The SCI-SLM team realised that it was premature to confidently speak about sustained management of the selected initiatives within the project lifespan as not enough experience had been gathered. In fact, most of the countries, with the exception of Uganda, have put the greatest effort and gained more insights into the initial steps of the methodology (identification, selection, characterisation and cross-visits), this focus being primarily justified by time and resource constraints. However, the evidence collected this far (see Chapters 5 to 8) provides a promising indication of the innovation path that the selected initiatives may follow.

Programme development processes

This component of the methodology was also covered in detail in Chapter 3. In brief, it illustrates how a project to promote farmer innovation can be established and managed, leading to the ultimate goal of embedding it into the existing research and extension system of a country, thereby (i) expanding the sources of innovation; and (ii) improving communication between farmers, scientists and field agents (Critchley 2007). Its graphic outline is reported in Figure 11.3 for ease of reference. As shown in the figure, the programme development processes, which should accompany field-based activities, represent building blocks in a pyramid graph with side arrows symbolising the gradual vertical upscaling. These processes reinforce each other and pave the way towards the final goal of institutionalisation. As for the field activities methodology, in this chapter, we will revisit the programme development processes with the objective of highlighting the most significant lessons and common denominators that can be derived from the experience of the four SCI-SLM countries.

Capacity building

The SCI-SLM executing organisation (CEAD-UKZN), the implementing agency (UNEP-GEF) and the Technical Advisory Group (CIS-VU) agreed that the capacity-building activities conducted during the early stages of the project, mostly through TAG training sessions and field visits, were crucial in ensuring that all country teams had a common understanding of the methodology and technical concepts from the outset. Further to this, the TAG, with its regular country visits, played a central role in overseeing and bridging the methodological experiences, thus maintaining agreement on the common lines throughout the project. At the

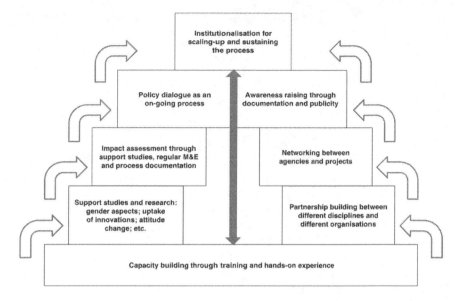

Figure 11.3 Programme Development Processes used to guide both PFI and SCI-SLM – as visualized under SCI-SLM

country level, each team developed its own capacity-building strategy. In Morocco, for instance, capacity building was carried out for both TARGA-Aide staff (SCI-SLM lead agency) and members of the innovative communities through the organisation of local workshops. It was noted that this was done for each of TARGA-Aide's projects and was consistent with the organisation's general mandate to build capacity for rural development. In Ghana, capacity building was primarily directed at the project staff of the University for Development Studies and the partner organisations (also members of the SCI-SLM national Steering Committee), with the additional goal of nurturing existing and potential partnerships.

Inventory of related programmes and initiatives, partnership building and networking

In practice these two elements of the programme development processes merged naturally and seamlessly. Experience showed that the type of lead agency (university, NGO, government) taking the initiative influenced the level, composition and robustness of the partnerships developed.

In Ghana, the University for Development Studies made a strategic choice in selecting the members of the national Steering Committee at the project's onset. The selected members were all key persons for the process of partnership building and, ultimately, institutionalisation. They represented government ministries and affiliated agencies, a consortium of NGOs, local NGOs and CBOs and a government-sponsored, community-based development project. The experience with governmental and non-governmental organisations was particularly positive as these were established institutions already involved in SLM policies, programmes and practices.

TARGA-Aide, in Morocco, supported the selected innovative communities in setting up partnerships with relevant administrative bodies such as the Department of Water Affairs and Forestry, the Department of Agriculture and the local administration *commune* (see Chapter 6). The team focused on expanding its network, for example, by presenting the SCI-SLM experience at the launch of the A2DTRM project (Support for Dynamic Development of Rural Areas in the Mediterranean) held in Rabat, Morocco, in February 2011.

In Uganda, the Ministry of Agriculture, Animal Industries and Fisheries forged strategic partnerships with relevant agencies with the additional aim of supporting data collection for M&E (e.g. partnership with the Department of Meteorology to access weather data relevant for the SCI-SLM project sites).

The Centre for Environment, Agriculture and Development of the University of KwaZulu-Natal in South Africa established links with different government agencies, such as the Department of Environmental Affairs and the Department of Agriculture, Forestry and Fisheries. Thanks to this collaboration, communities developed their understanding of the relevant policy framework and legislation and were thus able to improve their (technical) capacity (see Chapter 7).

Support studies

Several support studies were carried out as part of or in cooperation with the SCI-SLM project with direct involvement of researchers, university students and other project partners. The support studies contributed to the project by: (i) providing a deeper understanding and knowledge of specific (socio-economic and institutional) aspects of the initiatives to the SCI-SLM team and the communities themselves (e.g. the forest initiatives in South Africa – see Chapter 7); (ii) supplementing the baseline information gathered in the characterisation forms with required, higher-quality data; (iii) addressing the research gaps on purely technical aspects of the innovations as part of the joint experimentation process (e.g. the Ugandan studies regarding the use of arboreal termites to control terrestrial termites and the use of the night *kraaling* system to rehabilitate degraded rangelands – see Chapter 8); and (iv) establishing valuable relations between the communities and relevant project partners. Uganda, in particular, benefited significantly from the pre-established student internship programme with the Centre for International Cooperation of the Vrije Universiteit Amsterdam (CIS-VU). Over the years, the Centre, in its capacity of backstopping team for the PFI, PROLINNOVA and more recently the SCI-SLM project, created a mutually beneficial match between students and projects' research needs. By way of example, Table 11.3 provides a synthetic overview of the main support studies carried out in Uganda.

Impact assessment

In all four countries, the impact of farmer-to-farmer exchange visits has been remarkable, as demonstrated by high rates of adoption and horizontal spread of new

Table 11.3 Support studies by students carried out under the auspices of SCI-SLM Uganda

Name of student	University	Thesis title/topic	Country of field work	Year of thesis submission
Mats van Amen (MSc)	Vrije Universiteit Amsterdam (VU)	Community-induced sustainable land management: potential for exchange of knowledge? (RECPA community)	Uganda	2004
Guido Spanjaard (MSc)	VU	Identification of community initiatives in sustainable land management: RECPA	Uganda	2004
Matthew Kitchen (MA)	University of Amsterdam (UvA)	An analysis of NGO/local farmer partnerships promoting indigenous knowledge and local innovation in Kabale and Tororo districts, Uganda	Uganda	2005
Mariya Deren (MSc)	VU	Local adaptation strategies to land degradation, RECPA	Uganda	2006
Melanie Vaessen (MSc)	VU	A study on the RECPA initiative	Uganda	2006
Angela Tejada Chavez (MSc)	VU	United we stand? Tree planting by a CBO (RECPA)	Uganda	2008
Zsofia Bossanyi (MSc)	VU	Gender and participatory innovation development in Uganda	Uganda	2009
Eva Laura van de Ven (MSc)	VU	Communities taking the lead – the relevance of social innovation to improve agriculture and enhance livelihoods amongst land-based communities: Banyakabungo community in Ntungamo district and BANDERA 2000 group in Kamuli	Uganda	2011
Olaf Piers (MSc)	VU	The diffusion of innovations in SLM: perceptions of smallholder land users in Uganda	Uganda	2011

Source: SCI-SLM monitoring data.

SLM initiatives (see Chapters 5–8). This may be associated with the practical field level interactions on technologies and practices and the free and open flow of information among farmers. The impact on the ground is particularly visible in the case of Uganda. By way of example, some impact data regarding the Bandera community are presented below:

- A total of 17 Bandera community members visited Soroti and Pallisa districts to learn about management of citrus orchards, and 14 members visited the NACIA community in Nakasongola district to learn about water harvesting. As a result of these visits, five Bandera members doubled their areas under citrus orchards and 21 farmers improved in-field SLM practices in their orchards, including water retention basins, organic manure application and mulching.
- Bandera community members involved other interested parties (including their neighbours, friends, relatives and local associations) in setting up and monitoring the conservation agriculture (CA) demonstration plots. Through this effort, at least 27 of the group members who established demonstration plots reached out to a minimum of three other community members who, in turn, took up the practice.
- The demonstrated results of CA in Bandera created increasing demand from local leaders to roll out the technology to other farmers. To meet this demand, the Bandera community established an outreach training team consisting of five members (three men and two women). The team targeted other interested community-based organisations and local leaders, as well as medium- and large-scale farmers within a radius of 30 km, offering field practical training in various CA practices. The team has so far reached out to 315 farmers in Kisozi, Nawanyago, Kitayundwa and Balawoli sub-counties in Kamuli district, and Irongo sub-county in Luuka district, where they have carried out 31 training sessions and facilitated the setting up of 108 CA gardens.
- In addition, other SLM projects have used the Bandera's elaborate in-field SLM demonstration centre to train their farmers. For example, 180 farmers from 14 groups under the SLM mainstreaming project in the districts of Kamuli, Kaliro, Nakaseke, Nakasongola, Sembabule and Lyantonde have visited Bandera community, particularly to learn CA practices.

Awareness raising

In addition to the general SCI-SLM brochure, the national teams produced leaflets, posters and other dissemination material for wide distribution to different audiences. The team from Ghana and Morocco were also very proactive in awareness raising among members of the scientific community by publishing articles in journals and specialised magazines. Footage material was also collected in Ghana and Morocco capturing key moments of project implementation (e.g. exchange visits, capacity-building, etc.) with the final goal of developing short awareness-raising videos. These videos are still work in progress. They go beyond the project's official commitments and represent a form of legacy for the future sustainability of SCI-SLM principles, methodology and approach.

However, it should be noted that, in the context of SCI-SLM, awareness raising was not the sole prerogative of the project team. The innovative communities played an important role. For instance, the Bandera community proudly shared its successful experience with conservation agriculture on radio and television. Since 2012, Bandera members have featured more than 15 times on programmes hosted by local radio stations (Radio Star, Basoga Baino, Nile Broadcasting Service and Kiira Radio). In addition, at least seven articles in *The New Vision* and *The Daily Monitor* newspapers featured Bandera community activities. Furthermore, in June 2012, Bandera community hosted the national activities to mark the World Day to Combat Desertification under the theme 'Healthy soil sustains your health'. The three-day event attracted over 3,000 participants who witnessed exhibitions and demonstrations on SLM practices such as use of the field soil-testing kit, plant clinic, biochar technology, conservation agriculture, agroforestry, manure composting, energy efficient fuel stoves, solar technology, agro-processing and meteorological weather monitoring, among others (see Chapter 8).

Policy dialogue

In this area, a very useful lesson was learnt from the experience in Ghana, where the national team made the strategic choice to invite as members of the project Steering Committee people who were directly influential at national (and local) policy levels. These individuals, who fully aligned with the principles and objectives of SCI-SLM, acted as ambassadors for the project and actively lobbied for the institutionalisation of its approach with some successful results in terms of vertical upscaling. Among the key messages which they brought into the policy dialogue, they stressed the need for the government, research establishments and NGOs to revise their largely top-down research and extension agendas especially in the area of agriculture and the environment and look towards effective incorporation of bottom-up methodologies. Top-down research and extension systems do not pay attention to very practical and more sustainable local community and individual initiatives (see Chapter 5).

Institutionalisation

Three out of the four SCI-SLM countries, Ghana, Morocco and South Africa, worked towards institutionalisation from different starting points and with different results (Uganda is an exception as will be explained later). None of these three national teams can confidently assert to have achieved widespread institutionalisation of the concept and methodological approach to local innovation within the project lifespan.

In Ghana, for instance, the University for Development Studies has achieved some degree of mainstreaming of SCI-SLM practices through the Ministry of Food and Agriculture in the Kassena-Nankanna District of the Upper East Region and through the Zasilari Ecological Farms Project in the West Mamprusi District of the Northern Region (see Chapter 5). TARGA-Aide, the lead agency in Morocco, faced major challenges as, for an NGO, even if well established and recognised at national level, it is far more difficult to promote institutionalisation. However, the fact that the SCI-SLM approach is fully integrated in TARGA-Aide's standard protocol of intervention and that the NGO continues to support the selected communities as part of its long-term rural development programmes ensures a certain level of sustainability (see Chapter 6).

Uganda is an exception on two main accounts: (i) the vertical upscaling was consistently stimulated by an agency (MAAIF) embedded in the Government system; and (ii) over the course of three consecutive projects (PFI, PROLINNOVA and SCI-SLM) enough time had elapsed and sufficient evidence had been collected for the local innovation approach to be fully embraced and become a common practice in Uganda's official research and extension system. Since 2011, under the oversight of the Ministry of Agriculture, Animal Industry and Fisheries, the SCI-SLM methodology has been adopted by the larger UNDP-DDC-funded sustainable land management programme in Uganda. The SCI-SLM national co-ordinator also manages the overall SLM programme. All SLM projects are overseen by the same Steering Committee. Further to this, all participating districts have an SLM co-ordinator supported by an SLM Task Force to keep all projects on the same track. In particular, SCI-SLM has influenced Uganda's programme on conservation agriculture, setting a successful example and standards for other CA projects (see Chapter 8). What is abundantly clear is that SCI-SLM has been able to catalyse activities under the national SLM programme in Uganda, particularly through (i) establishing community entry points for the focus of activities; and (ii) stressing the importance of community-to-community exchange of information. Peer learning through exchange visits has become routine and helps adoption of new innovations. The two programmes are intertwined and there is a mutually beneficial organic relationship. SCI-SLM provides the entry point, broad methodology and 'community as trainers'; and the SLM programme takes this forward through its network, funds and implementational capacity, thus creating a win-win relationship with promising future developments.

Finally, it should be noted that, even though the country experiences with institutionalisation differ on many points, there is a shared view of the fact that mainstreaming of the SCI-SLM approach has been based on visible successful demonstration of practices by the communities themselves and tangible evidence of their results and benefits. Therefore, any achievements in terms of institutionalisation carry a substantial weight.

Concluding remarks

The SCI-SLM methodology has proved to be a useful tool in stimulating community initiatives towards both vertical and horizontal upscaling. It was not written in stone and, in general, allowed the level of flexibility and participation needed to make such a 'community innovation' approach work, at least in its basic form, in different geographic and institutional contexts. The general lesson learnt from its practical implementation is that the potential of local innovations can be fully 'exploited' only when:

- innovators become aware of the value and importance of their innovations and are willing to share their experience with others;
- the innovation is recognised as relevant by many potential users;
- there is an enabling/conducive environment for the stimulation and spread of the innovations; and
- local and scientific knowledge merge (hybrid knowledge) in a synergetic and constructive way for the further improvement of the innovation.

As noted in the project's external mid-term review, the potential for upscaling innovations is high using the SCI-SLM methodology, as shown in Uganda, where SCI-SLM principles, approach and methodology have been embedded within the national sustainable land management programme and community initiatives have started spreading on their own, beyond MAAIF's facilitation (Kojwang 2013). The Uganda experience confirms that government structures and policy support are crucial in any attempt to upscale operations. However, while working through government (central level) has advantages in terms of institutionalisation and sustainability, it has also drawbacks, especially in terms of bureaucratic procedures and associated costs. Therefore, as confirmed by the experience of all four SCI-SLM countries, the central government cannot operate in isolation. It is important to establish genuine partnerships and links with local government institutions and other stakeholders (e.g. NGOs, CBOs, academic institutions, research centres, etc.) that can act more nimbly and move resources quickly enough when needed.

It can be said that SCI-SLM, despite the fact that it set out to refine its methodology, has actually changed little in its initial methodological approach. This is for two main reasons. First, on a positive note, this is because the methodology has served well as a basic framework and each country has found it a useful guide to working with community innovators – interpreting each step as it best fitted that country. Second, it is apparent, as the project comes towards its close, that relatively little analytical thought has been invested in the details of methodological development. Process documentation, that is, the recording of experiences both at country and programme levels, could also have been more systematic. This is hardly surprising as the day-to-day implementation of such a project is very demanding, and pragmatic decisions, rather than abstract reflections, demand priority attention.

Notes

1   In collaboration with: Saa Dittoh and Conrad Weobong, University for Development Studies, Ghana; Mohamed Mahdi and Zakaria Tijani, Targa-Aide, Morocco; Avrashka Sahadeva, Michael Malinga and Maxwell Mudhara, University of KwaZulu-Natal, South Africa; Stephen Muwaya, Ministry of Agriculture, Animal Industries and Fisheries, Uganda; Richard Molo and John Ssendawula, Makerere University, Uganda; Mohamed F. Sessay, UNEP, Kenya.

2   Promoting Farmer Innovation (PFI) was a Dutch Government-funded, UNDP-UNSO coordinated project active in Kenya, Tanzania and Uganda from 1997 to 2001. Technical assistance was provided by the Vrije Universiteit Amsterdam's Centre for International Cooperation.

210   *Di Prima, Critchley and van de Ven*

References

Critchley, W. (2007), *Working with Farmer Innovators – A practical guide*. CTA.

Critchley, W. and Di Prima, S. (2009) SCI-SLM Methodology guidelines. Leaflet prepared by the Technical Advisory Group (TAG) for the project inception meeting in September 2009; last revision: 13/09/2010.

Di Prima, S. and Critchley, W. (2011), Preliminary findings from baseline analysis. SCI-SLM working paper.

Di Prima, S., Critchley, W. and Tuijp, W. (2013), *Methodology for Working with Innovative Communities – The SCI-SLM experience*. UNEP-GEF policy brief.

Kojwang, O.H. (2013), *Mid Term Review of the Project 'SCI-SLM'*. UNEP.

SCI-SLM, (2009), SCI-SLM Inception Workshop in South Africa: Proceedings report.

SCI-SLM, (2013), *Survey on the Evolution of SCI-SLM Methodology and Approach at Country and Programme Level*. SCI-SLM book-writing workshop, South Africa.

UNEP, (2009), *Stimulating Community Initiatives in Sustainable Land Management (SCI-SLM)*. Project document.

van de Ven, E. (2011), *Communities Taking the Lead – the relevance of social innovation amongst land-based communities in Uganda to improve agriculture and enhance livelihoods*. Master's thesis in Environment and Resource Management, Vrije Universiteit Amsterdam.

12 Lessons learnt and conclusions

William Critchley, Sabina Di Prima, Maxwell Mudhara and Saa Dittoh

We bring this book to an end by summarising what we have learnt over the course of the SCI-SLM project. Naturally this is from an insider's point of view: all of those involved in this book are 'insiders' – and cannot claim to be independent. But that does not detract from the value of the observations, as our perspective is unique and stems from a longstanding relationship with the project. This final chapter looks at the impact of SCI-SLM. What broad lessons have emerged most clearly to inform future initiatives? While the final word, and more quantitative analysis, will be the prerogative of the official Final Evaluation team, they will do well to bear in mind what is concluded here.

South-to-South learning

During the preparation phase, a visit to Morocco brought a very clearly expressed wish: 'we would like to know more about what our fellow Africans are doing, rather than only receiving advice from Western experts'. Morocco had been chosen to complete the four nations representing the different regions of Africa. But perhaps it had not been adequately realised just how North Africa was almost totally cut off from its Sub-Saharan neighbours by geography, and to a large extent by language, too. South-to-South learning was high on the agenda of SCI-SLM: this served to reinforce it. During the course of the project it transpired that communication and cross-learning between countries was enthusiastically embraced by all. While impossible to quantify, there was nothing but positive attitudes. A willingness to learn was matched by keenness to offer advice. Perhaps it was surprising to all that so many challenges and opportunities were shared, at least broadly, across the four corners of the continent.

There were many fascinating encounters. For example, community members from the Atlas Mountains of Morocco visiting northern Ghana could immediately connect with local problems of soil fertility – and themselves make suggestions on rangeland management to a pastoral group in Uganda. Translators were available, but proved not always to be as vital to communication as might have been thought. Though expensive to move groups to other, often distant, parts of the continent, the rewards may be correspondingly great. It is quite simple to organise regional transfer and, though this is increasingly common, it still happens too rarely. No wonder there is increasing attention being given to bespoke 'learning routes' within Africa. SCI-SLM's experience wholeheartedly supports this.

Within-country exchange

Within-country exchange is equally part of South-to-South learning. SCI-SLM certainly took that view, and part of its methodology was to encourage inter-community exchange. African communities are extremely insular. This was carried out by each country, in many different ways: much was achieved and, while over the whole project this particular activity was somewhat independently organised, it was unanimously considered effective. Hosting groups not only displayed their innovations, but also welcomed ideas from visitors – and built relationships. Recognition empowers, and appreciative visitors endow communities with greater ambition and determination. Positive feedback is always a prime motivator.

In Ghana two groups independently involved in compost making, and situated not far from each other (about 80 km), were put in contact by the project. They immediately engaged in enthusiastic discussions about their initiatives, on both technical – and social – matters, and both came away with new ideas. Uganda had perhaps the most systematic programme of cross-visits (or 'exchange visits') and certainly the best documented impacts (see Chapter 8). While groups with similar initiatives made for the most obvious exchange visits, others with quite different initiatives proved to be very interested in one another. Just because one group is involved with grazing management doesn't mean the members have no lessons to exchange with fruit growers. It must be remembered that innovative social organisation was as important as technical SLM initiatives under SCI-SLM – so, again, this was a topic of interest to visitors.

Study visits were used by South Africa and Uganda, in particular. In South Africa, groups were ferried to research stations and commercial organisations to great effect: those managing wattle trees were taken to forest stations to understand 'scientific approaches' to forest management (Chapter 7). In Uganda, the BANDERA community were transported to other districts to learn about fruit tree management, agroforestry and conservation agriculture (Chapter 8).

Community as entry point

Why was SCI-SLM's emphasis on communities as an entry point to innovation and learning in SLM? Part of the rationale was that projects such as Promoting Farmer Innovation (see Chapter 2) had previously concentrated their attention on individuals, yet quite often found that communities were also involved in innovation. Furthermore, designing a project to look at community initiatives opens the door to *social* innovation rather than just *technical* innovation. Thus, a new angle is shed on how people in Africa address problems of land degradation: and that is through social cohesion. Reorganisation of communities may be the answer to help achieve tangible and technical objectives (see Chapter 3 for SCI-SLM's working definition of 'community'). Crucially, this focus also means that a further spectrum of SLM initiatives are brought into play: those that address common property resources of grazing land, irrigation schemes, wetlands and forests.

While community initiatives may imply an equal and joint effort, it quickly became clear that central to any community initiative is (usually) a charismatic individual. Meetings with any of the groupings inevitably drew out the ring leader, the man or woman who had inspired the others and brought a team together. Even when there is a dominant personality, in contrast to projects that home-in specifically on individuals, the danger of the 'favoured farmer' syndrome is much reduced. Potential jealousies associated with projects lavishing attention on a particular farmer are dissipated to a great extent when a community is the focus.

Any doubts that innovative communities would not be uncovered were quickly dispelled. No country had a problem in the identification stage, and it became clear that SCI-SLM's initial hypothesis was well founded. Naturally the 'communities' themselves varied, with examples of whole villages in Uganda, sub-village communities in South Africa, common interest groups in Morocco and relatively small self-help (women's) groups in Ghana. Community-based innovation turned out to be a response to a variety of factors, some predictable, some not: problems associated with common resources such as grazing arrangement (Uganda), or the need for land reclamation (Morocco, Uganda) and better fertility management (Ghana) were predictable, but insecurity regarding cattle theft, or shortage of 'burial logs' (both South Africa) were not.

A final note concerns working with existing, self-initiated and largely self-sustained communities. Consistently, SCI-SLM teams found communities responsive and willing. Not only do they constitute pre-formed groups – rather than ones swiftly assembled to tap into project resources – but they demonstrate stability and a sense of common purpose. In many countries in Africa they can register themselves as self-help associations and be receptacles for further support; also, communities represent a cost-effective way of delivering extension and other forms of assistance.

Local technical innovation

SCI-SLM put community innovation under the microscope in terms of both technologies and social arrangements. These sometime occurred side by side in a single community. The innovative technologies in SLM that were identified as valid (according to the clear SCI-SLM criteria: see Chapter 3) ranged widely. However they were bound by a common denominator in that they were valuable as responses to important local problems of land degradation – and the potential of SLM to address those challenges. Many of the technical SLM achievements uncovered are those that development agencies strive to find answers for: intractable problems where researchers often fear to tread. Six main categories stand out.

- **Grazing management:** *'turning the tragedy of the commons into common management'*
  In both South Africa and Uganda there are examples of communities recently taking over management of common lands, when the prevailing development wisdom is that traditional forms of community management have broken down and have no future. In these two countries SCI-SLM found that this need not necessarily be the case (Chapters 7 and 8 respectively).
- **Forest management:** *'taking over the forest from the foresters'*
  Management of forests in Africa is plagued by difficulties and uneasy relationships

between forest departments and local communities. Just as participatory forest management has made an impact in Asia and Latin America, so there are green shoots appearing Africa. SCI-SLM showcases examples of community initiatives in Morocco, South Africa and Uganda (Chapters 6, 7 and 8, respectively).

- **Invasive alien species management:** *'changing alien invasives into local assets'*
  South Africa presents not just one, but two, examples of communities taking control of black wattle (*Acacia mearnsii*) copses and, instead of seeking to eliminate this invasive alien (a thirsty, introduced species from Australia, still grown in estates), managing the trees and making productive use of them, simultaneously challenging official policies of eradication (Chapter 7).

- **Non-burning in dry season:** *'breaking tradition for a better future'*
  Innovation usually builds on, or modifies, tradition. In Ghana, burning of grass and residues post-harvest has been practised traditionally to clear land and flush out edible fauna for the pot. In Chapter 5, cases are presented where communities themselves have banned this practice for their own reasons. This is entirely in line with development recommendations (especially with climate change concerns), but directly initiated by communities here.

- **Soil fertility management:** *'organic recycling: gender roles turned round'*
  Not only are two separate groups in Ghana involved in compost making to improve soil fertility and health, but they are women's groups too: and that challenges local gender roles. This, traditionally, is men's work. Here the technical aspect of the initiative is sound and creative, but the social arrangement even more significant in its innovativeness (Chapter 5).

- **Rehabilitation of degraded land:** *'productivity restored: regreening with a purpose'*
  Land rehabilitation is increasingly a concern in a world where agriculture is running out of options for expansion. Costs are generally high, and examples of rural groups achieving this themselves through novel methods, without outside prompting or material support, is simply unheard of. But in Morocco and Uganda this is exactly what is happening. Significantly, in both cases, the trigger was more land for productive purposes, not reducing degradation for the global good (Chapter 6 and 8 respectively).

In terms of the technologies adopted, the vast majority of the local innovations were, unsurprisingly, low-external-input interventions. One valuable implication of this is that their potential for spread and chances of sustainability are enhanced. Finally, on a technical point, while the selected SLM initiatives were basically sound, all offered the chance of betterment in one or more aspects of the TEES-test (see following section on methodology). We have already noted that exchange visits led to ideas for improvements from other communities. The more conventional approach to upgrading innovation is through the 'participatory innovation development' approach espoused by PROLINNOVA (see www.prolinnova.net). However 'joint experimentation', which brings together researchers, extension workers and local people, proved hard to achieve, just as it has in other projects. Uganda does, however, present an exception (Chapter 8). The theory is sound, but practice faces several obstacles: not least being the reluctance of researchers to amend their professional agendas and priorities.

Social innovation

In many ways the articulation of 'social innovation' was one of the main strengths (and indeed innovative aspects) of SCI-SLM itself. Too often there is a fixation with technological advancement, rather than social arrangements for progress. Conventional development wisdom holds that traditional community cohesion is coming unstuck. Perhaps it may be in many cases, but SCI-SLM found plenty of examples of new social arrangements formed for a common purpose. It has already been noted that, by looking specifically at community, rather than individual, innovation, SCI-SLM was able to embrace social organisation. And it was also able to recognise it as an innovation in its own right, when associated with an SLM technology that was deemed to be – at least – good practice (see Chapter 3).

Many groups – indeed most – were composed of both women and men. Sometimes exclusive women's groups (e.g. in Ghana) were found to be active in SLM. A basic common denominator was working together to solve similar problems. SCI-SLM had initially embraced a methodology that characterised social innovation on the basis of the SRI-test (Sustainable; Replicable; Inclusive) but, after a dedicated support study, found that more appropriate was the 'SERR' test, which looked for the four elements of Sustainability, Efficiency, Replicability and Responsibility. These then, were the criteria that SCI-SLM considered should be used in a future search for genuine social innovation.

Social innovation, it was broadly agreed, is more difficult to identify but also to spread to other communities than technical innovation because it is intangible, and more difficult to define. Furthermore, it is more community, and culture, specific.

Methodology

Two chapters of this book are dedicated to SCI-SLM's methodology and its evolution (Chapters 3 and 11). This surely is one of the key contributions of SCI-SLM: a meticulously thought-out methodology based on experience with prior, pioneering projects – and further analysis in this book of how it evolved. With respect to selection of initiatives, we have noted above the criteria test for social innovation and how it developed over the course of the project from the SRI-test to the SERR test. The TEES-test for technical innovation proved simpler and readily adopted: here the acronym TEES, based on 'Technically effective, Economically valid, Environmentally friendly and Socially acceptable' works, conveniently, both in French as well as English.

Notwithstanding training/capacity building, there was, however, a degree of confusion caused by the initial stages of the methodology. This is more to do with the definition of an innovation/initiative (used synonymously under SCI-SLM) and the difference between *social* and *technical* innovation (and the fact that a social innovation does not need to have an innovative SLM technique associated with it, but must at least be associated with sound SLM practices). Thus, in order to make sure that SCI-SLM stayed true to purpose, and focused on 'true' innovation, there was a need to filter out what was simply 'good practice' or 'research recommendation', otherwise spontaneous innovation disappears out of view. Clear criteria were required – and provided.

Other aspects of the methodology, both field activities and programme development processes, were less complex (see Chapters 3 and 11). Though not all the stages were reached in all countries, this did not prevent the broad methodology being accepted as a useful guideline. It was accepted that flexibility must be embraced. After all, each country partner had pre-existing approaches which were mixed and matched with SCI-SLM.

Spread of Innovation

An analysis carried out two years after initiation of the project is reported in Chapter 11: yes, 'horizontal spread' of initiatives certainly happens spontaneously, and SCI-SLM has surely stimulated this. Unfortunately (see next section on monitoring and evaluation) there was no later analysis of spread, something that the final evaluation of the project should surely carry out as part of its work. Extremely important, though impossible to quantify, is the spread of 'innovativeness'. The ultimate aim of sustainability is much more likely to be achieved if change in behaviour has roots in attitudes within local communities and support from agencies working with them: this could be said to represent SCI-SLM's 'Theory of Change', even though it was never articulated as such.

It was widely agreed by SCI-SLM partners that intra-country exchange visits between communities promoted by the project were the main route for spread of technical innovations and to a certain extent social innovation. This was backed up by an equally strong feeling that land users are more likely to adopt from one another than from an official extension agent. After all, the route 'practitioner-to-practitioner' has been the main source of spread for millennia. Where adoption has occurred, this is genuine, because it is voluntary, without pressure and without material inducement. The role of regional visits under the project has not been assessed: but testimonies from those who have travelled and exchanged ideas at regional meetings, hosted in turn by each participating country, gives the strong impression that ideas and inspiration have travelled across the continent (Chapter 9).

Where there remains a paucity of evidence is 'vertical spread' or institutionalisation. This is the pinnacle, literally, of the SCI-SLM methodological process, where it sits at the apex of the graphic (see Figure 11.3 in Chapter 11). Nevertheless, Uganda's national SLM programme has adopted a large part of the SCI-SLM methodology and has taken communities to be its main entry point for extension work (Chapter 8). This has occurred mainly as a result of the fact that the national SCI-SLM coordinator is simultaneously the national SLM programme coordinator. In other countries, at least all coordinating agencies have been deeply influenced by the approach, and additionally each country has a Steering Committee that has representation from various in-line ministries. It is certainly to be hoped that the GEF – and UNEP too – will reap lessons from SCI-SLM that can be integrated into their respective work programmes.

Monitoring and evaluation

Monitoring and evaluation has been relatively weak. This is unsurprising as it is, sadly, a common malady of such development projects. Despite a carefully designed documentation process, based on forms that had been developed under Promoting Farmer Innovation, it is only the initial characterisation of initiatives – an analytical stocktaking exercise – that was comprehensively completed. While meeting official requirements for narrative and financial reporting, follow-up monitoring of what has happened in the field has been irregular and haphazard. There are exceptions though – Uganda presents quite comprehensive data in Chapter 8.

This is not an issue that SCI-SLM alone has faced: it is the fate of most similar projects. There appears to be little appetite for strong and systematic M&E, because it is time-consuming (competing with other more 'constructive' work) and there is little reward and seldom adequate positive feedback. Surely one answer would be better supported programmes of M&E, with dedicated budget lines and specific staff allocated. The mid-term evaluation of SCI-SLM told us little that we didn't already know: the final evaluation is an opportunity to pick up vital data. An assessment of impact, and recommendations for replication of specific aspects of the methodology into other projects, and other countries, is a priority of that exercise.

Where now?

Despite its limited period of field activity (the fate of too many development projects) SCI-SLM itself has surely been vindicated in its hypotheses regarding the abundance of relevant community innovation in SLM. The communities uncovered are, unwittingly, following the new creed of SLM with its acknowledged links to poverty reduction, livelihoods, ecosystem function, climate change adaptation and mitigation. However, neither the communities themselves, nor development agencies generally, realise just how important these local innovations are, and how powerful is the undercurrent of creativity.

Certainly the project's approach has struck a chord with the current development emphases on communities, innovation, resilience, landscape approaches and cross-learning – so there are lessons here that should not be ignored. These lessons can help build donor–researcher partnerships that could take the current SCI-SLM initiative to the next level beyond its practical, developmental achievements and its undoubted 'intellectual inspiration' into formally recognised and publicly supported innovation systems.

The focus of the project is on localised, community-driven innovations in sustainable land management under changing environmental conditions. This places it firmly in line with the priorities of the United Nations Convention to Combat Desertification, the TerrAfrica initiative and the Global Environment Facility – together with a host of other development agencies. Thus the expectation is that lessons rooted in methodological, technical, socio-economic and institutional issues can, and should, be readily applied outside the project areas and be used to shape the future directions in the programmes under the UNCCD, TerrAfrica and the GEF in Africa.

Are most of the answers to Africa's land degradation problems to be found *within* the continent? And are a significant proportion of these the result of local innovation processes? Certainly a lack of travel opportunities, and often language barriers too, are major constraints to dissemination of knowledge and spread of ideas. Surely all SLM-themed projects should make room for carefully targeted exchange visits and also keep an eye open for innovation, both technical and social.

Metaphorically speaking SCI-SLM exposes low-hanging fruit, ripe for harvest, which has not yet been spotted. However, a caveat: the emphases on South-to-South learning with a community focus, and on local innovation in SLM (both technical and social), does not imply that this is presented as an alternative paradigm for development in Africa. Rather, these are areas that certainly deserve more recognition and can add vital supplementary value to current approaches.

Index